TRANSACTIONS OF THE

AMERICAN PHILOSOPHICAL SOCIETY

HELD AT PHILADELPHIA

FOR PROMOTING USEFUL KNOWLEDGE

———————

VOLUME 68, PART 4 · 1978

The Science of Minerals in the Age of Jefferson

JOHN C. GREENE

PROFESSOR OF HISTORY, UNIVERSITY OF CONNECTICUT

and

JOHN G. BURKE

PROFESSOR OF HISTORY, UNIVERSITY OF CALIFORNIA, LOS ANGELES

THE AMERICAN PHILOSOPHICAL SOCIETY

INDEPENDENCE SQUARE: PHILADELPHIA

July 1978

Library of Congress Catalog
Card Number
International Standard Book Number 0-87169-
US ISSN 0065-9746

PREFACE

Meeting for the first time in 1967 at a symposium on the history of geology, the authors discovered a common interest in the history of mineralogy. One of us, Professor Greene, was interested in the activities of a group of devotees of mineralogy—physicians, college teachers, immigrant naturalists, gentlemen of leisure—in the early American republic. The other, Professor Burke, had done research in the history of mineralogy and published *Origins of the Science of Crystals* in 1966. We decided to combine our interests and information in describing the early development of mineralogical studies in the United States in relation to the emergence of scientific mineralogy in Europe in the years from 1780 to 1822.

This book is the outcome of that collaboration. In it we attempt to create the scientific scene in the early American nation through the activities, aspirations, and achievements of a relatively small band of mineralogists and to show the relation between their efforts and accomplishments and those of the more advanced scientific community in Europe. At the same time, we reveal the theoretical structure and practical methodology in that period and sketch in the background of previous developments in mineralogical science.

The authors are indebted to several people and institutions for assistance in completing this work. They are especially obliged to Professors Clifford Frondel of Harvard University and Norman Gray of the University of Connecticut for criticizing the manuscript and assisting in the technical evaluation of American mineralogical writing and of Adam Seybert's collection of minerals. They are also grateful to Professors Edmund Spieker, Marland Billings, Janet Aitken, Paul Farrington, and John T. Wasson for their suggestions with respect to particular chapters and to George W. White for his careful reading of the entire manuscript. In the matter of financial assistance in undertaking and completing the research for this book, the authors wish to acknowledge their indebtedness to the National Science Foundation, the American Philosophical Society, the John Simon Guggenheim Memorial Foundation, the Research Foundation of the University of Connecticut, and the Research Committee of the Academic Senate, University of California, Los Angeles.

In quoting from manuscript materials, the authors have taken the liberty of modernizing the punctuation whenever it seemed desirable from the standpoint of readability.

The history of science is a vast and complicated field of study, and the history of early American science presents special problems in maintaining a balanced perspective with respect to the importance of American contributions in the total scientific enterprise. It is hoped that the present work will show the possibility of overcoming these obstacles through collaborative research and writing.

JOHN C. GREENE,
University of Connecticut
JOHN G. BURKE,
University of California, Los Angeles

3

THE SCIENCE OF MINERALS IN THE AGE OF JEFFERSON

JOHN C. GREENE and JOHN G. BURKE

CONTENTS

I. MINERALOGY BECOMES A SCIENCE

Selecting a precise date or period when the study of a particular aspect of nature becomes a science is not an easy task. Mineralogy is one of the most ancient fields of human interest. Middens from the paleolithic and early neolithic periods yield crystals of quartz, serpentine, jasper, and lapis lazuli, showing that prehistoric man collected and prized these semi-precious stones. Archaeological investigations and tomb paintings reveal that in the three millennia before the time of Aristotle artisans had identified such mineral ores as cuprite, malachite, galena, cassiterite, stibnite, and hematite. Yet, the first known classification of minerals is that of Aristotle's colleague Theophrastus (370–287 B.C.). His primitive arrangement separated mineral substances into three categories: those subject to the force of fire, those that were incombustible, and those that were soluble in water and vitrifiable by heat.[1] This scheme, with variations and gradations, lasted well into the eighteenth century.

As early as the middle of the seventeenth century, however, students of minerals had acquired sufficient knowledge and clarity of purpose to begin using special words to identify themselves and the substances they studied. The word *mineral* has a Celtic root and entered the Latin vocabulary about the twelfth century. Thereafter it rapidly replaced the older Latin word *metallum* in designating metallic ores, contrasting with the word *lapis*, which was used to identify silicates and rocks. The term *mineralogist* seems to have appeared first in English in Sir Thomas Browne's *Pseudodoxia Epidemica: or Enquiries Into Very Many Received Tenets, and Commonly Presumed Truths,* published in London in 1646. Referring to the popular belief that rock crystal was permanently

concreted water, Browne argued that many of the "exactest mineralogists" denied this.[2] As for the word *mineralogy,* the earliest use of this term appears to have been in 1690, in Robert Boyle's essay *A Previous Hydrostatical Way of Estimating Ores.* According to Boyle:

. . . our impractised way of estimating ores, may not be useless; and for that reason will not perhaps be unwelcome to some that love mineralogy, much better than they understand it; especially coming forth at a time when many industrious persons of this nation are excited to look after profitable minerals. . . .[3]

Boyle called practitioners of mineralogy "mineralists," a name that came to be used much less frequently than Browne's term "mineralogist."

Despite these early developments, the establishment of mineralogy as a science must be placed in the decade of the 1780's, the period that saw the publication of René Just Haüy's *Essai d'une théorie sur la structure des crystaux* (1784), Lavoisier's *Traité élémentaire de chimie* (1789), and the *Méthode de nomenclature chimique* of Lavoisier, Guyton de Morveau, Berthollet, and Fourcroy (1787). The exploration of the chemical, physical, and crystallographic properties of minerals received a great impetus from these works, and the knowledge thus gained resulted in the elaboration of improved schemes of classification in the early years of the nineteenth century.

A number of sociological and psychological influences combined to aid in the establishment of the science of mineralogy. Boyle's remarks serve to illuminate the most important stimulus to the development of mineralogy, namely, the interest in the economic exploitation of minerals for the metals they contained, for the medicines that might be prepared from them, or for their value as gem stones. Although by present standards the quantities of metals produced in the rural civilizations of Europe and colonial America were insignificant, metals were nevertheless employed in a multitude of products. Iron hollow ware—pots, pans, and kettles—was cast directly from blast furnaces, as were anchors, fireplace andirons, and pokers. In addition to their trade in horseshoes, blacksmiths dealt in wire nails, wheel rims, scythes, sickles, hay rakes, knives, saw sets, razors, shears, axes, chains, and plow irons. Coppersmiths sold kettles,

---

[1] *Theophrastus's History of Stones,* John Hill, tr. (London, 1746).

[2] Sir Thomas Browne, *Pseudodoxia Epidemica* (London, 1650), p. 37.

[3] Robert Boyle, *The Works of the Honourable Robert Boyle,* Thomas Birch, ed. (6 v., London, 1772) 5: pp. 489–490.

boilers, and roofing copper as well as stills for the manufacture of beer and whiskey. Tin was alloyed with copper to produce bronze, which found major uses in statuary and bells; tinned iron plates were formed into pipe and water spouts. Speculum metal for use in mirrors was manufactured from an alloy of copper and tin with a small amount of arsenic or antimony; tin hardened with minor additions of copper, bismuth, or antimony provided pewter tableware. Lead was employed in ribs for window panes, for roofing, as sheathing for ships, and as type metal when alloyed with antimony and bismuth. The precious metals, gold and silver, were used as coinage, in jewelry manufacture, in *objêts d'art,* and for tableware in the homes of the wealthy. And, as armaments became larger and more sophisticated, increasing quantities of the basic metals—iron and copper—went into brass and iron cannon, and iron and steel were employed to manufacture swords, muskets, bayonets, pikes, and shot.

The Romans worked mines of various metallic ores in Europe and Britain: gold in the Basque area of northern Spain and at Dolaucothy in southern Wales; silver at Cartagena in Spain; copper at Río Tinto in southwestern Spain; lead at Mendip in Britain; tin in northern Spain, in Brittany, and in Cornwall and Devon; and iron in eastern Austria, in the Weald and the Forest of Dean in Britain, in Luxembourg and northern Alsace. From the tenth century forward, these Roman mines were reopened and many other ore locations discovered. Prolific silver deposits, for example, were exploited at Guadalcanal in Spain, at Kongsberg in Norway, and particularly in Bohemia and Saxony, and the vast iron ore deposits of Bergslagen in central Sweden commenced to be worked systematically.

In the sixteenth century Vannoccio Birunguccio, Georgius Agricola, and Lazarus Ercker wrote treatises describing the extraction, assaying and production of metallic ores, which faithfully mirror this increased economic interest in minerals.[4] In the seventeenth and eighteenth centuries the desire to produce metals efficiently and economically led to the establishment of government sponsored mining schools in several European countries. Sweden was the first nation to recognize the value of these technical institutions. In 1684 the Swedish government placed the chemical laboratory of Urban Hiärne under the jurisdiction of the Board of Mines and gave it the task of training stu-

dents for careers in the metals industries. Many Swedish scientists who later distinguished themselves in analytical chemistry, mineralogy, and metallurgy received their training at the Academy, among them Emanuel Swedenborg, Sven Rinman, Axel Cronstedt, and Johann Gahn.[5]

Another early school was the Mining Academy at Selmecbánya (presently the city of Banská Stiavnica in Czechoslovakia) in the kingdom of Hungary. Established in 1735, it offered education in the theory and practice of mining and metallurgy and attracted students from all European countries. Among the faculty were Nicolas Jacquin, a native of Leyden, who served as professor of practical mining and chemistry from 1763 to 1769; Johannes Scopoli, subsequently a professor at Pavia; and Antal Ruprecht, a former student of the Swedish scientist Torbern Bergman. Four students of the Academy distinguished themselves especially in their subsequent careers. In 1801 Andrés Manuel del Río announced the discovery of a new element which he called erythronium; rediscovered in 1831, it was given its modern name vanadium. The Spanish brothers Juan José and Fausto d'Elhuyer first isolated tungsten in 1781, and Ferenc Mueller von Reichenstein in 1782 suspected the presence of a new element, tellurium, in gold minerals; it was chemically extracted by Martin Klaproth in 1798.[6]

The Bergakademie established at Freiberg in Saxony in 1766 became one of the leading institutions of this kind. Originally intended to educate only inhabitants of Saxony, it opened its doors almost immediately to students from other German states and, in 1771, to those of foreign countries. Entering classes usually numbered about twenty students, about one-third of whom were foreign. As with the Swedish Academy, many of the Bergakademie graduates later ranked among the most respected mineralogists and geologists in Europe—Leopold von Buch, Alexander von Humboldt, Friedrich Mohs, Dietrich Karsten, and Christian Samuel Weiss. The Bergakademie international prestige was due in large measure to the presence on its faculty from 1775 until 1817 of Abraham Gottlob Werner, renowned both for his system of mineral identification and classification by external characters and for his steadfast adherence to the theory that water was the chief agent of geological change.[7]

In France the government prepared for the establishment of a mining academy in 1769 by imposing a special tax on persons holding mining concessions from the Crown. But it was not until 1778 that Balthazar Sage was appointed to a chair of mineralogy

[4] Vannoccio Biringuccio, *De la Pirotechnia libri X* (Venice, 1540) has been translated into English by Cyril Stanley Smith and Martha T. Gnudi (New York, 1943); Agricola, *De Re Metallica Libri XII* (Basel, 1556) has been translated into English by Herbert Hoover and Lou H. Hoover (London, 1921); and Lazarus Ercker, *Beschreibung allerfürnemsten mineralischen Ertzt-und-Berckwerksarten* (Prague, 1574) has been translated into English by Annaliese G. Sisco and Cyril Stanley Smith (Chicago, 1951).

[5] Theodore A. Wertime, *The Coming of the Age of Steel* (Chicago, 1962), p. 259.

[6] Ferenc Szabadváry, *History of Analytical Chemistry* (London, 1966), pp. 45, 81.

[7] *Festschrift zur hundertjährigen Jubiläum der Bergakademie zu Freiberg* (Dresden, 1866), *passim.*

and assaying at the Paris mint nor until 1783 that the École des Mines was formally organized by the addition of another chair. The duration of study was three years. The students pursued their academic studies in the winter and learned the practical aspects of mining in the field during the summer. In 1790, after the outbreak of the Revolution, the school was closed. It reopened in 1794 when the Bureau (later Council) of Mines was created. Classes were composed of thirty students, twenty of whom had completed the curriculum of the École Polytechnique. In 1802 Napoleon closed the school at Paris, establishing two other schools devoted to the practical aspects of mining and mineral exploitation, one at Geislautern in the Saar basin for training in the mining of iron ores and coal, the other at Moutiers in Savoy for instruction in the mining and metallurgy of nonferrous metals. The schools were closed in 1814 when the French lost these territories, and the École des Mines was reopened in Paris in 1816. Almost all of the leading mineralogists and geologists in post-Revolutionary France were graduates of the École des Mines, among them André Brochant de Villiers, Élie de Beaumont, Louis Cordier, and Armand Dufrénoy.

The influence of these schools on the development of mineralogical science can scarcely be exaggerated. Their success is demonstrated by the rapid rise in the production of minerals in Europe in the nineteenth century and by the exploitation of the mineral resources of European colonial empires around the globe. Their combined practical and theoretical curricula produced graduates who made important contributions both to mining technology and metallurgical practices and to the sciences of geology, mineralogy, and crystallography.

The government of Great Britain did not sponsor a mining academy similar to those in Europe. All of the continental governments had more or less direct ownership or control of mineral deposits and had, therefore, an interest in seeing to it that minerals were mined efficiently and economically. This situation did not obtain in Great Britain. There were, of course, professors who taught geology and mineralogy at Cambridge and Edinburgh and sister institutions, and Richard Kirwan attempted unsuccessfully about 1800 to establish a mining academy at Dublin similar to that at Freiberg. But those students who wished to pursue further their scientific or practical interests traveled to Europe and enrolled usually either at Freiberg or the École des Mines. The United States followed the English example; institutions whose curricula were devoted to mining and metallurgy did not exist until the Missouri School of Mines was established at Rolla in 1870 and the Colorado School of Mines at Golden in 1874.

Minerals were also sought and studied for their medicinal value, either real or imagined. Classical works and medieval encyclopedias, particularly the lapidaries, abound in information about the preventative and curative powers of certain minerals. Supposedly the amethyst precluded intoxication; the heliotrope and diamond guarded the body against poisons; hematite cured certain diseases of the blood; and selenite was a remedy for diseases of the chest. Experimentation to prepare medicines or pharmaceuticals from metals or minerals stemmed primarily from Paracelsus, who taught a new theory of disease in the early sixteenth century. Greek and Roman physicians had believed that disease consisted in an imbalance of four humors within the body. But Paracelsus considered disease to be a disturbance of the body's chemistry and drew an analogy between the chemical basis of various functions of the human body and certain natural processes such as the combination of minerals in the earth and the growth of plants. He and his followers placed chemistry in the service of medicine, using the distillations or residues of plants and minerals as specific pharmaceuticals. Medical recipe books containing Paracelsian remedies became common in the late sixteenth century, and these treatises culminated in the publication of the great *Pharmacopoeia Londinensis* in 1618.[8] Owing to the primitive state of the sciences of chemistry and human physiology, few of the medicines were efficacious; nevertheless, this activity did result in increasing knowledge of chemical reactions and the chemical constituents of minerals.

Since antiquity mineral waters have been enjoyed for their restorative powers or prescribed internally as a medicine. Baden-Baden in Germany and Baden-bei-Wien near Vienna were famous for their refreshing mineral springs in Roman times, as were many spas in Roman Britain and Gaul. There were attempts in the classical period to discover the substances to which mineral waters owed their creative properties, and these efforts multiplied during the Middle Ages and the Renaissance. Such studies contributed to the development of chemical analysis. By the seventeenth century reports of the results of these tests had become common in the literature of science. Qualitative tests or analyses were gradually superseded by quantitative analyses, in which the combining weights of the ingredients were calculated without any thought of a theory or mechanism of chemical combination. In the process, a number of the chemical elements, then termed "simple bodies," began to be isolated.

The search for precious gems and the attempts to explain their remarkable properties provided another stimulus to the development of mineralogy. Even in prehistoric times gemstones were valued for their intrinsic beauty and rarity; gems have been found in

---

[8] Allen Debus, *The English Paracelsians* (London, 1965), pp. 32–35, 49, 152–156.

middens of the pre-dynastic period in Egypt and of the third millennium B.C. in India. In the Roman period, Pliny the Elder described many gemstones in his *Natural History,* commenting on their rarity and beauty. The most influential early modern work describing gemstones and mineral crystals was by Anselm Boéce de Boodt, a physician and curator of the mineralogical collection of Emperor Rudolph II. His *Gemmarum et Lapidum Historia* was first published in Prague in 1609; its popularity resulted in additional Latin editions and translations. De Boodt gave detailed descriptions of over six hundred mineral substances, faithfully relating legends about their marvellous properties, the majority of which he discounted. He also attempted to clarify some of the properties of gemstones qualitatively.

De Boodt's work provided a model for later natural philosophers who studied and tried to explain the various properties of gemstones and mineral crystals: their colors, the action of heat and light upon them, their cleavage, and particularly their regular geometric figures. Most authors explained colors as resulting from the accidental entrainment of metallic ingredients, a supposition later proved to be correct only with respect to some minerals. Franz Ulrich Aepinus, a professor of astronomy in Berlin, conducted the first systematic investigations in tourmaline in 1756 of the production of electrical polarity in crystals with a temperature increase (pyroelectricity), although he remarked in his report that this phenomenon had long been known to jewelers.[9] Erasmus Bartholinus in 1669 first described the double refraction of light in crystals, subjecting the Iceland crystal, calcite, to a thorough study.[10] Robert Boyle, believing that mineral crystals were composed of thin extended lamellar plates, attributed the directional character of cleavage to a natural tendency for parting to occur parallel to these layers.[11] And Nicolaus Steno in 1669, by speculating on the manner in which mineral crystals were formed in the earth, led the way in the exploration of the reasons why they exhibit regular geometric figures.[12] These studies resulted in the genesis of the science of crystallography in the late eighteenth century.

Mineralogical collections among titled and well-to-do amateurs increased sharply during the eighteenth century, providing yet another stimulus to the development of the science of mineralogy. Anselm de Boodt was curator of Emperor Rudolph II's mineralogical collection. But the passion for collecting was not confined to emperors. In Paris alone the number of known mineral cabinets increased from seventeen in 1740 to sixty-one in 1780. Collectors exchanged minerals to diversify their collections, and the nobility exchanged rare specimens as gifts. In 1780 Archduchess Marie-Anne of Austria, who had formed an extensive cabinet, presented several beautiful specimens of the tourmaline of Tyrol to Duke Charles of Lorraine to enrich his collection.[13]

As collecting increased so did the desire to arrange these collections according to some standard scheme. As the eighteenth century progressed, wealthy amateurs called upon natural philosophers to arrange their collections in some rational order. In 1767 Jean-Baptiste-Louis Romé de l'Isle, a founder of the science of crystallography, prepared a three-volume catalog describing systematically the magnificent natural history collection of Pedro Franco Davila, a wealthy Peruvian who decided to auction his holdings. In his personal copy of the catalog Romé de l'Isle entered the purchase price, and in many cases the purchaser, for each lot of the collection, from which Davila realized approximately $228,000 in current money. The geologist Nicolas Desmarest, the mineralogists Balthazar Sage and Louis Daubenton, the noted architects François Belanger and Jean-Marie Morel, and the famous physiocrats Turgot and Quesnay, all attended the auction and made purchases. The titles attached to many other names show that the auction provoked great interest among the nobility of Parisian society.[14]

There are numerous other examples of the arrangement of mineral cabinets of the wealthy by early mineralogists. Count de Bournon, an émigré in England during the Revolution and the Napoleonic period, organized the large mineral collections of Lord Grenville, Sir Abraham Hume, and Sir John Aubyn. After the restoration of the Bourbons, Bournon returned to France and became director general of the mineralogical cabinet of Louis XVIII. Friedrich Mohs, upon completion of his education at the Academy at Freiberg, became curator of the mineral collection of J. F. von der Null, a Viennese banker. Later Mohs moved to Graz, where he was curator of the mineral

[9] Franz Aepinus, "Mémoire concernant quelques nouvelles expériences électriques remarquables," *Histoire de l'Académie Royale des Sciences et Belles Lettres de Berlin* 12 (1756): p. 105.

[10] Erasmus Bartholinus, *Experimenta crystalli islandici disdiaclasti quibus mira et insolita refractio detigitur* (Hafniae, 1669).

[11] Robert Boyle, *An Essay about the Origine and Virtues of Gems* (London, 1672), pp. 22–23.

[12] Nicolaus Steno, *De solido intra solidum naturaliter contento dissertationis prodromus* (Florence, 1669).

[13] Arthur Birembaut, "La Minéralogie," in: M. Daumas, ed., *Histoire de la Science* (Paris, 1957), p. 1074. See also *Journal de Physique* 15 (1780): p. 196 fn.

[14] Information obtained from Romé de l'Isle's copy of the work, now in the private library of John G. Burke. Jean Louis Marie Daubenton (1716–1800), anatomical illustrator for the Count de Buffon's *Histoire Naturelle, Générale et Particulière* and curator of the natural history collections of the Royal Cabinet for fifty years, was attracted to mineralogy through his interest in paleontology. He published his *Tableaux méthodique des minéraux* in 1784 and was named professor of mineralogy in 1793.

cabinet of the Johanneum for several years. Appointments such as these gave the men who received them the opportunity and incentive to develop logical systems of classifying the minerals under their care. More often than not, these mineralogist-curators published the results of their work, so that classification became an integral part of the science of mineralogy.

To sum up: The exploitation of mineral ores for metals or medicinal ingredients, the interest in geometrically shaped gemstones for their monetary or aesthetic value or for their unusual properties of doubly refracting light or becoming electrically polarized when heated, the search for the source of the curative powers of mineral waters, and the passion for collecting minerals focused mineralogical studies in three major areas: the chemical analysis of minerals, the structure and physical properties of mineral crystals, and the rational classification of mineral substances. The progressive acquisition of knowledge in these areas jointly contributed to the founding of the science of mineralogy.

CHEMICAL ANALYSIS

One of the earliest analytical tests on record, mentioned in the records of the ancient kingdoms of Mesopotamia and Egypt and in the Old Testament, is the fire-assay. In this operation, performed to control the purity of gold and silver or to detect the counterfeiting of gold or silver coinage, the piece of metal is weighed before and after prolonged heating, during which substances added for the purpose of counterfeiting are separated from the pure gold or silver. The more difficult task of detecting the mixing of gold with silver was first accomplished by Archimedes of Syracuse, who resorted to quantitative methods. By measuring the amount of water displaced by masses of pure gold and pure silver equal in weight to Hiero's crown, Archimedes was able to determine that the goldsmith who had made the crown had substituted some silver for the gold Hiero had given him for its manufacture.[15]

The touchstone, or "Lydian stone," was another early semi-quantitative method for testing the purity of gold. Mentioned by Theophrastus, it consists in comparing a standard color sample with the color of the streak produced by rubbing the metal sample against the stone, usually a fine and even grained black jasper, slate, or igneous rock. The touchstone test was greatly refined during the Roman period, when comparison was made with a number of needles, or pieces of known composition. When copper began to be added to gold and silver coinage, the composition of the needles was similarly altered. By the sixteenth century the number of standard needles employed in the test had multiplied, and in skilled hands the

composition of alloys of gold and silver could be measured in the touchstone test to one part in one hundred.

Pliny the Elder reported the only wet analytical test known in antiquity. It involved the use of an extract of nutgalls to detect the adulteration of verdigris or copper acetate, which was used in salves to remedy eye ailments and ulcers of the skin. Because of its cost, verdigris was subject to adulteration by the addition of green vitriol, or ferrous sulfate. By addition of nutgalls caused the adulterated mixture to turn black; thus the test was actually an indication of the presence of iron, although this was unknown at the time.[16]

In the medieval period two trends aided the development of analytical chemistry: the practice of alchemy and the qualitative analysis of mineral waters. The most important contribution of the alchemists was the discovery of nitric, hydrochloric, and sulfuric acids, which eventually enabled assayers and chemists to dissolve metals. The origin of the processes necessary to prepare these acids is still a matter of conjecture. Tenth-century alchemical works describe rudimentary procedures; in the late thirteenth century the preparation of nitric acid was clearly delineated; and by the mid-sixteenth century the preparation and use of the acids in mining and metallurgy was common. The achievement was largely due to improvements in distillation equipment, which enabled alchemists to handle corrosive substances at high temperature, and to the introduction of condensate cooling.[17]

In the same era some progress was made in the qualitative analysis of mineral waters. The ancients were aware that rain water is pure, that the waters of springs contain various mineral ingredients depending upon the locality and the types of surrounding strata, and that boiling improves the quality of water. No systematic efforts to determine the ingredients of mineral waters, however, are recorded before the Middle Ages. The early tests consisted in evaporating a sample of water and examining the residue by taste and smell. Gradually other tests were introduced: the measurement of specific gravity, the detection of iron by the addition of nutgalls, and the identification of some of the ingredients in the boiled or distilled residue by color or characteristic crystal figure.[18]

Robert Boyle, the seventeenth-century English scientist, introduced the systematic use of extracts of

[15] *Vitruvius: The Ten Books of Architecture*, M. H. Morgan, tr. (New York, 1960), Book IX: pp. 253–254.

[16] *Theophrastus's History of Stones*, pp. 78–80; Agricola, *De Re Metallica*, pp. 253–260, 439–441; K. C. Bailey, *The Elder Pliny's Chapters on Chemical Subjects* (2 v., London, 1929 and 1932) 2: pp. 41, 172.

[17] Szabadváry, *History of Analytical Chemistry*, p. 13; Robert P. Multhauf, *The Origins of Chemistry* (London, 1966), pp. 140, 162–163, 173, 204.

[18] Allen G. Debus, "Solution Analysis Prior to Robert Boyle," *Chymia* 8 (1962): p. 43; also Debus, *English Paracelsians*, pp. 158–160. Szabadváry, *History of Analytical Chemistry*, pp. 29–30.

violets and cornflowers, cochineal, and litmus solutions for testing the acidity or alkalinity of mineral waters. He defined the process of precipitation and used flame tests to insure the identification of copper, sulfur, silver, and sodium compounds. Boyle's work in general analytical chemistry has been rightly seen as a landmark. His biographer Marie Boas writes:

> He could test for acids, alkalis, most common metals, certain salts and acids; and in his use of specific gravity and crystal form he had a method of differentiating unknown substances, if not of identifying them. And he was continually extending his range by devising new tests and new methods.[19]

A number of chemical tests may be made rapidly and easily on minerals by the use of a blowpipe. The simple blowpipe is a tapering iron tube ending in a small opening through which air from the lungs can be forced under a relatively high pressure. When this current of air is directed into a flame, combustion takes place rapidly and produces sufficient heat to melt or fuse many minerals. Fusible materials can be classified according to the ease with which they fuse, and with suitable flames and fluxes the constituents can be identified by flame color, odor, or residue.

Johann Kunckel suggested the employment of the blowpipe in 1679, and its use grew slowly during the early eighteenth century. In Sweden, at the hands of Axel Cronstedt and Torbern Bergman, the blowpipe became a standard mineralogical instrument. Cronstedt introduced the use of sodium carbonate, borax, and sodium phosphate as fusion mixtures. Bergman, in 1779, published his *Commentario de tubo ferrumintario,* in which he described methods of obtaining a suitable flame, the various fusion mixtures and their proper use, and the reactions of various mineral compounds when subjected to blowpipe analysis.[20]

Even before the theoretical achievements of Lavoisier and his colleagues in the 1780's, chemists had begun to supplement qualitative tests with quantitative analyses, that is, to determine not only what the constituents of a substance were, but also their relative quantities. Further, they began to achieve the ability to distinguish between the various compounds of a metal. By carefully reproducing experiments without any clear theoretical basis, they came to the realization that the mixture of certain reagents would result in the formation of certain compounds and that the original reagents could be recovered by additional chemical processes. The accumulation of knowledge about chemical combinations, particularly the recognition that the ingredients of air and other gases might also be involved, eventually provided a basis for Lavoisier's theoretical advance and later for that of Dalton.

One of the first to publish a quantitative result was Nicholas Lémery, who in the latter part of the seventeenth century reported the combining weight of silver in silver chloride within an accuracy of 4 per cent. A greater impetus toward the use of the chemical balance came from the experiments of Joseph Black in 1754 on the carbonates of calcium and magnesium, in which he isolated carbon dioxide gas and realized that this gas was held in combined form in carbonates. Black's reliance on the balance made his results incontrovertible, and his discovery of a combined gas stimulated others to experiment in similar ways. By 1780 a total of ten new gases had been isolated: carbon dioxide, carbon monoxide, ammonia, chlorine, oxygen, nitrogen, sulfur dioxide, and the three oxides of nitrogen.

As a byproduct of his work Black also differentiated hydroxides and carbonates. About the same time other chemists were discriminating other similar substances, such as sodium and potassium salts. S. A. Margraff, for example, compared sodium nitrate and saltpeter, noting that the crystals of the two substances differed and that they yielded different colored flames. Converting the nitrates to sulfates, chlorides, and carbonates, Margraff detected differences in the resulting crystals and in the solubility of the corresponding compounds, thus clearly establishing the distinction between the compounds of the two elements.

As in blowpipe analysis, Torbern Bergman led the way in establishing the basic methodology of analytical chemistry. In several treatises collected and published as *Opuscula physica et chemica* (6 vols., Uppsala, 1779–1790) Bergman described his techniques and showed their applicability to the analysis of mineral waters and mineral ores. He explained how samples should be prepared, how solvents should be purified and acids used; he recommended the use of distilled water, of glass vessels, and the repeated washing of precipitates; and he delineated the correct procedure for the weighing of precipitates. Bergman also recorded the colors of the metal precipitates and presented tables of combining weights of a number of substances, which, though inaccurate in many instances, set the pattern for subsequent analytical texts.[21]

Bergman's work was translated almost immediately into German, French, and English. In the three following decades his procedures were refined and extended, and his combining weights were corrected by a number of chemists, the most influential of whom were Martin Klaproth in Berlin and Louis Vauquelin in Paris.

In the United States, in the early years of the nineteenth century, the most widely used textbook of chemical analysis was Thomas Thomson's *A System of Chemistry* (4 vols., Edinburgh, 1802). Thomson

---

[19] Marie Boas, *Robert Boyle and XVIIth Century Chemistry* (Cambridge, 1958), pp. 139, 108–141.
[20] Szabadváry, *History of Analytical Chemistry,* pp. 51–54.
[21] *Ibid.,* pp. 58–59, 71–81, 86–88.

was intimately familiar with all of the contemporary literature of analytical chemistry. His recommended methods and techniques for the analysis of minerals closely follow those of Klaproth and Vauquelin. For the analysis of mineral waters he depended upon Richard Kirwan's *An Essay on the Analysis of Mineral Waters* (1799).

Thomson's text demonstrates the level of sophistication of analytical chemistry in 1800. The lacunae and the pitfalls in his procedures from the modern point of view are apparent. He was vague about the strength of concentration of the reagents and omitted any consideration of the presence of zinc in his analysis of silicates. He did not know the components of hydrochloric acid; he was aware of the two ferric oxides, Fe_2O_3 and Fe_3O_4, but not ferrous oxide, FeO; and although he recognized four oxides of lead, Pb_2O_3 had not yet been isolated. Yet, considering that analytical chemists about 1800 lacked any clearly stated theory of chemical combination and failed to recognize the existence of a number of elements, it is remarkable that they attained the accuracy that they did. With some noticeable exceptions, Thomson's list of combining weights by modern standards is accurate within about 5 per cent.

By 1800, then, chemists had developed fairly reliable techniques for analyzing waters and minerals, and the urge toward quantification had resulted in the ability to determine the percentage quantities of the constitutents. In the case of simple ores such as sulfides or oxides of metals, the combination of the two elements was assumed. With the complex silicates the various oxides were isolated and their weights reported. It would be over a century before the correct chemical combinations of the silicates were fully understood. Nevertheless chemistry, through experimental analysis rather than by theory, had made a major contribution to the establishment of the science of mineralogy. Its further contributions would be in the perfection of experimental techniques, in the isolation and discovery of new elements in minerals, and in the more precise calculation of the percentages of mineral constituents by the use of atomic theory.

THE STRUCTURE AND PROPERTIES OF CRYSTALS

As early as the Renaissance the shapes of crystals were used as a means of identifying the ingredients of mineral waters. In a more sophisticated and systematic manner, by the seventeenth and eighteenth centuries, chemists learned to distinguish such compounds as sodium nitrate from potassium nitrate. The peculiar geometric shapes of crystals had been noted in classical antiquity, arousing curiosity as to the causes responsible. Discussing rock crystal Pliny the Elder wrote:

Why it is formed with hexagonal faces cannot readily be explained; and any explanation is complicated by the fact that, on the one hand, its terminal points are not symmetrical and that, on the other, its faces are so perfectly smooth that no craftmanship could achieve the same effect.[22]

Early explanations of crystal shapes, heavily influenced by Aristotelian thought and animistic and alchemical accretions to it, were couched in such organic and developmental terms as formative drives, organizing germs, and latent principles. In the seventeenth century, mechanical explanations were first suggested, and these finally resulted in the establishment of the science of crystallography by Haüy.

Both Johann Kepler and René Descartes toyed with mechanical models in their attempts to explain the shapes of snowflakes, but Robert Hooke extended the idea to include all crystalline substances. In his *Micrographia* (London, 1665), Hooke described how three identical, contiguous spheres would form an equilateral triangle, four a rhombus, five a trapezoid, and so forth, demonstrating that three-dimensional solids would also result from the stacking of such particles. In an attempt to explain the double refraction of light in the Iceland crystal (calcite), Christian Huygens outlined a view of the structure of crystals that was similar to that of Hooke with the exception that the contiguous particles were ellipsoids of rotation rather than spheres.[23]

Although Hooke and Huygens were atomists, their ideas were not accepted because they could not be integrated into the prevailing theory of matter and because they contradicted empirical evidence. A major tenet of the seventeenth-century atomic theory was that atoms were qualitatively homogeneous, that is, formed of one universal matter. The distinctive qualities of substances had to be explained on other grounds, and the shape of the atoms was considered to be the primary differentiating factor. If all atoms were identical spheres, as Hooke theorized, and each particle had exactly the same qualities, as was generally believed, all macroscopic substances, or at the very least all crystals, should be identical in odor, taste, smell, and touch. Further, in all the microscopic studies of Leeuwenhoek, Freind, Bourguet, and others, the tiniest bits of matter observed were angular rather than spherical; hence the atoms composing the bits were presumed to be angular as well. The idea that crystalline shapes might be explained by the stacking of spherical atoms fell on barren ground. It was not until the early nineteenth century that William Hyde Wollaston revived the theory.

[22] *Pliny: Natural History,* H. Rackham and W. H. S. Jones, tr. (Cambridge, Mass., 1938) 37: p. 26.
[23] Johann Kepler, *Strena seu nive sexangula* (Francofurti ad Moenum, 1611); Charles Adam and Paul Tannery, eds., *Œuvres de Descartes* (12 v., Paris, 1897–1913) 6: p. 288; Robert Hooke, *Micrographia* (London, 1665), pp. 85–86; Christian Huygens, *Traité de la lumière* (Leiden, 1690).

During the eighteenth century, the view that crystals were composed of polyhedral molecules gained rapidly in popularity, but the idea had one major flaw. Several substances crystallized in more than one configuration; calcite, for example, was found as rhombohedra, as hexagonal prisms, and as scalenohedra. For years mineralogists failed to explain why and how a substance composed of identical polyhedral molecules could crystallize in such a variety of shapes.

Torbern Bergman, in 1773, was the first to propose a solution to this problem. If one subjects a specimen of calcite to cleavage, there results a rhombohedron, whose surfaces have angles of approximately $101\frac{1}{3}°$ and $78\frac{1}{2}°$. Bergman suggested that if one took a large rhombohedron of calcite and placed upon each face successively smaller layers the various shapes of calcite would result, depending upon the amount of the reduction in size of the successively applied layers. By this method Bergman successfully produced a common configuration of calcite consisting of an hexagonal prism terminated at each end by rhombohedral faces as well as two scalenohedral forms. However, he was unable to reproduce the hexagonal prismatic form of calcite terminated at each end by flat plane surfaces. Wiser observers, Bergman concluded, were required to bring the hypothesis to fruition.[24]

Bergman's hope was soon fulfilled. Within a decade René Just Haüy had solved the problem. On several occasions Haüy asserted that he was completely unacquainted with Bergman's work until he had informed the French Academy of Sciences of the substance of his first two memoirs in 1781. In a recent article, however, R. Hooykas argued that Haüy was completely in Bergman's debt, since in his first two memoirs Haüy remained on the same qualitative level as Bergman. Between 1781 and 1783, Haüy arrived at the idea that the superimposed layers were composed of cleavage rhombohedra, and then developed his laws of decrement, which formed the basis of his entire system.[25]

To account for the variety of crystal shapes, Haüy first had to determine by the cleavage of a substance the shape of the basic building block of the substance and the value of its plane and interfacial angles.[26] Haüy called the resulting piece the constituent molecule or the nucleus, and found that the nuclei of all of the crystals he studied had one of six different shapes or primitive forms. The primitive forms, however, could be mathematically subdivided and hence reduced to three, which Haüy termed integrant

molecules. They were the parallelepipedon, the triangular prism, and the tetrahedron. For each substance that he studied, Haüy determined by geometry the values of the plane and interfacial angles of the integrant molecule.

Haüy taught that a crystal is formed by the progressive addition of layers to each face of the nucleus, each layer consisting of contiguous integrant molecules and having the thickness of one molecule. His first law of decrement involved the successive subtraction of integrant molecules from all of the edges of the newly added layer. For example, assuming that the nucleus of the crystal is a cube and that the added layers are composed of molecules that are also cubes, if one row of molecules is subtracted from the edges of each successively applied layer, the cube will eventually take the shape of a dodecahedron with rhombic faces (fig. 1). If the growth of the crystal ceases before a perfect dodecahedron is completed, the crystal will have a cubic shape, but in the place of the edges there will be elongated hexagons included at angles of 45° to the faces. Haüy explained that the minuteness of the molecules precluded even microscopic observation of the stairlike aspect of the successively smaller layers.

In some instances, Haüy found it necessary to postulate differential decrement; for example, there might be a decrement of two rows of cubic molecules on two opposite edges of each successively deposited layer with a decrement of only one row of molecules on every other new layer on the other two edges. Such a process produces a dodecahedron with pentagonal faces. Because the molecules were units, Haüy could calculate by the use of plane trigonometry the plane angles of the sides of the faces as well as the interfacial angles of the resulting crystal.

Another law of decrement operated on the angles or parallel to the diagonals of the primitive form. By this law an octahedron could be generated from a cube. In addition to these fundamental laws there could be other intermediate decrements. For example, the superimposed layers might be decreased by two rows of molecules at the edges and at the same time by triple the thickness of the simple molecule. Sometimes, the decrement operated on certain sides and certain angles and not on others.

Haüy did not consider crystallization as a dynamic process. He did not speculate as to why differential decrement occurred. He only wished to demonstrate the capability of identical integrant molecules giving birth to a multitude of different forms. The purpose of crystallography, he said, was to determine the mathematical relationships existing between the nucleus or primary form and any secondary form. There was a connection between the secondary crystal forms of each species and the primitive form, and the work of the crystallographer entailed the determination

[24] Torbern Bergman, *Physical and Chemical Essays*, Edmund Cullen, tr. (3 v., London, 1784) **2**: pp. 9–23.

[25] Reijer Hooykas, "Les débuts de la théorie crystallographique de R. J. Haüy d'après les documents originaux," *Revue d'Histoire des Sciences* **8** (1955): pp. 319–337.

[26] For a detailed exposition of Haüy's work, see John G. Burke, *Origins of the Science of Crystals* (Berkeley and Los Angeles, 1966).

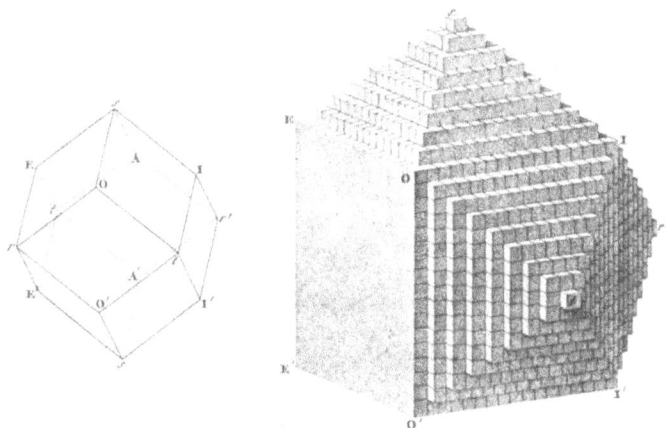

FIG. 1. Haüy's illustration of how the dodecahedron with rhombic faces is constructed by the progressive decrement of one row of molecules on each edge of layers successively added to a cubic nucleus. Source: R. J. Haüy, *Traité de minéralogie* (5 v., Paris, 1801) 5: pl. II, figs. 11, 13.

of the laws which linked the corresponding parts. Haüy considered his theory to be complete; by following his precepts and method, the unknown nuclei of all crystalline substances could eventually be identified.

Haüy's work in crystallography was a major contribution to the development of mineralogy. His meticulous attention to the forms of crystals allowed him to unite in the same species minerals that had previously been listed as separate species, such as beryl and emerald, and to distinguish a number of minerals previously lumped together, such as the "zeolites." Above all, Haüy directed attention to the internal structure of crystals and demonstrated that regularity of the external configuration was merely a manifestation of internal order. Haüy's laws of decrement led directly to the development of modern systems of crystal classification and to the idea of the space lattice.

THE CLASSIFICATION OF MINERALS

The classification of any group of physical objects or other natural phenomena is only as successful or advanced as the state of knowledge about the objects or the phenomena classified. Attempts to classify minerals before the first quarter of the nineteenth century reflected their authors' philosophies of nature and, particularly, the state of contemporary theory in chemistry, physics, and crystallography.[27]

Theophrastus was apparently the first to attempt to classify minerals. His classification of rocks and minerals into three categories—those subject to the force of fire, incombustibles, and those soluble in water and

vitrifiable by heat—though modified, was hardly improved until the middle of the eighteenth century. The classification of Agricola in his *De natura fossilium* (Basel, 1546, revised in 1558) was more practical. Agricola described five classes: (1) earths, such as clay and ochre; (2) stones, including gems, semi-precious stones, and unusual minerals; (3) "solidified juices," such as salt, alum, and vitriol; (4) metals; and (5) compounds of mixtures and simple substances, consisting of such minerals as galena and pyrite. In his descriptions, Agricola referred to most of the external characters commonly noted today—friability, odor, taste, color, hardness, luster, etc.—and attributed the variation in these properties to the different proportions of the Aristotelian elements—air, earth, fire, and water.[28]

The morphological approach to nature was typical of eighteenth-century science, not only in mineralogy, but also in botany and zoology. The first major mineralogical work was Maurice Cappeller's *Prodromus crystallographiae, de crystallis impropriae sic dictis commentarium* (Lucerne, 1723), in which forty substances were classified under nine different shapes. Linnaeus, in his *Systema naturae* (Stockholm, 1768), classified minerals under four generic shapes: the hexagonal prism, cube, tetragonal prism, and octahedron.[29] The most important work of mineral classification in the eighteenth century, however, was that of J. B. L. Romé de l'Isle. His initial effort, *Essai de cristallographie* (Paris, 1772), was followed by a four-volume masterpiece, *Cristallographie, ou description des formes propes à tous les corps du règne minéral*

[27] For a more detailed description of mineral classification see John G. Burke, "Mineral Classification in the Early Nineteenth Century," in: Cecil J. Schneer, ed., *Toward a History of Geology* (Cambridge, Mass., 1969), pp. 62–77.

[28] Agricola, *De re metallica*, p. 594.
[29] Carolus Linneaus, *Systema naturae* (4 v., Holmiae, 1766–1768), 3: pp. 3–7, 84–103, 213–216.

(Paris, 1783). Romé de l'Isle classified about 450 minerals under six primitive forms: the regular tetrahedron, cube, regular octahedron, rhombohedron, rhombic octahedron, and dodecahedron with triangular faces. He attributed the variety of shapes which a substance might assume to a process of truncation which operated on the primitive form during the process of crystallization, and whose cause, he confessed, was unknown. His second work was most important, however, because of his application of goniometry, or the measurement of crystal angles. The constancy of the interfacial angles of a mineral crystal had been recognized in quartz by Nicolaus Steno in 1669 and must certainly have been known to such men as Hooke, Huygens, and Cappeller. This law, however, first clearly stated in Romé de l'Isle's work, was drawn to his attention by a student, Arnold Carangeot, the inventor of the contact goniometer.[30]

An entirely different approach to the study of minerals was by the description of the various external characters. The dominating figure in this area was Abraham G. Werner, a renowned professor of mineralogy at Freiberg. Werner developed the science of oryctognosy, or the determination of mineral species by the detailed examination and description of their external characters. Werner's system was described in a work, *Aeusserliche Kennzeichen der Fossilien* (Leipzig, 1774), which he wrote when he was only twenty-five. Although he promised to publish a second edition, none ever appeared under his authorship. Instead, revisions and improvements which he described in his lectures were incorporated in a French edition, published in 1790, and in an English edition, published in 1805.[31]

The result of Werner's work was not a mineralogical system in the ordinary sense; it was instead a system of mineral characters. Werner readily admitted that the classification of minerals and their identification by the study of their external characters were two different tasks. He firmly believed that it was not useful to attempt to catalog minerals into classes, orders, genera, and species. These were artificial divisions created by man; in nature there are only individuals.

Each mineral, Werner taught, had four essential characters: the external or sensible properties; the internal or chemical composition; the physical, including electrical and optical, properties; and the empirical characters, deduced from the place of its occurrence and from associated minerals. Though admitting the importance of chemical composition, Werner relegated chemistry to a secondary role, partially because analysis was somewhat unreliable but chiefly because of its inconvenience when studying minerals in the field. Instead, he stressed the study of external characters, teaching that they must be determined as accurately as possible and described with the same exactness of expression as prevailed in mathematics.

In his view there were seven different external characters: color, cohesion, unctuosity, coldness, weight, smell, and taste. All of the senses were brought into play. These were generic characters, and each could be divided into specific characters, which, in turn, might be further described in terms of varieties. He enumerated eight principal colors; white, gray, black, blue, green, yellow, red, and brown. Green had eight varieties, ranging from verdigris green—a rather brilliant bluish green, in which mixture yellow is not perceptible, to canary green—a rather light yellowish green, constituting the transition from the green color into yellow. Beyond distinguishing the varieties of green, exact expression required that the intensity of the color be noted, that is, whether it is dark, clear, light, or pale.

Werner subjected each of the generic characters similarly to divisions and subdivisions to allow every possible nuance of expression in describing the mineral. The manner of cohesion, in particular, Werner demonstrated, could yield a vast amount of information. An immediate division could be made between fluids, solids, and friables. Solids could then be examined with respect to external form, external surface, external luster, internal luster, fracture, the form of the fragments, transparency, streak, stain, hardness, solidity, flexibility, adhesion to the tongue, and sound. Of the four different types of fracture—compact, fibrous, striated, and foliated—a compact fracture might be described as scaly, even, conchoidal, uneven, or earthy.

Werner's system was attacked and derided by some mineralogists as unphilosophical, fit only for miners and uneducated people.[32] But its essential practicality and utility won it a place in all mineralogical textbooks. Richard Kirwan followed Werner's system faithfully in his *Elements of Mineralogy,* and Kirwan's work was in the knapsacks of the majority of American field mineralogists in the early nineteenth century.[33]

Werner's critics were chiefly adherents of the idea that minerals should be classified in the same manner as botanical specimens, that is, by classes, orders, genera, and species. The crystallographer René Just Haüy was a leader in this trend, and it comes as no surprise that he was an amateur botanist before studying crystallography and mineralogy. Though Haüy

[30] J. B. L. Romé de l'Isle, *Essai de cristallographie* (Paris, 1772); *Cristallographie, ou description des formes propres à tous les corps du règne minéral* (4 v., Paris, 1783).

[31] Abraham Gottlob Werner, *On the External Characters of Minerals,* Albert V. Carozzi, tr. (Urbana, Ill., 1962).

[32] Richard Chenevix, "Réflections sur quelques méthodes minéralogiques," *Annales de chimie et de physique* 65 (1808): pp. 5–43, 113–160, 225–277.

[33] Richard Kirwan, *Elements of Mineralogy* (2 v., London, 1794).

insisted that the integrant molecule of a mineral crystal was a specific character of any substance and could in many instances be used as a means of identification without recourse to chemical analysis, in his major classificatory work, *Traité de minéralogie* (Paris, 1801), he did employ chemical criteria for classification.

Haüy's system comprised four classes: (1) acidiferous substances; (2) earthy substances; (3) nonmetallic combustible substances; and (4) metallic substances. The orders and genera of the first class contained minerals in which the alkali and alkaline earth metals and members of the magnesium and aluminum family were combined with acids. Species of the first genus of the first order were calcite, apatite, fluorite, and gypsum; those of the second genus of the first order were barite and witherite; those of the third, celestite and strontianite; and those of the fourth, epsomite and boracite. Saltpeter was a unique species of the first genus of the second order; the species of the second genus were halite, borax, and sodium carbonate; sal ammoniac was a unique species of the third genus. The third order consisted of a unique genus containing two species, alum and cryolite.

Haüy's second class—earthy substances—consisted entirely of species having the common feature that no acid had been detected in their composition. It included a variety of minerals: quartz, such oxides as corundum, anatase, spinel, and chrysoberyl, and many of the silicates. Non-metallic substances, comprising Haüy's third class, consisted of two orders: simple and combined. The first included three—sulfur, diamond, and anthracite—while the second had five—bitumen, bituminous coal, jet, amber, and mellite. Haüy's fourth class, metallic substances, was composed of three orders: (1) those not readily oxidized; (2) those able to be oxidized and readily reduced; and (3) those able to be oxidized but not readily reduced. Genera of the first order were platinum, gold, and silver; mercury was the unique genus of the second order; seventeen known metals made up the genera of the third order.[34]

Although Haüy's classification scheme had many partisans, within a short time both mineralogists and chemists proposed new systems. One that gained widespread popularity was published in 1807 by Alexandre Brongniart, then professor of natural history at the École Central des Quatres-Nations and later Haüy's successor as professor of mineralogy at the Muséum National d'Histoire Naturelle.

Brongniart's second class was almost the same as the first class of Haüy, except that he eliminated the third order and its single genus and included alum and cryolite as a species of a genus in the order containing alkaline earths. His fourth class, combustibles, corresponded closely to Haüy's third class.

Brongniart's fifth class, minerals having a metallic base, was divided into two orders, fragile metals and ductile metals; it included three new elements that had been discovered in the six years separating his work from that of Haüy. Charles Hatchett in 1801 had found niobium, though it was called columbium by him and Brongniart. A. G. Ekeberg had discovered tantalum in 1802; and Martin Klaproth had isolated cerium in 1803.

One of the major deviations from the classification of Haüy made by Brongniart consisted in establishing a new class, his first, termed non-metallic oxides. This class comprised two orders. The first, non-acidic oxides, had as genera air and water. The second, acidic oxides, had as genera sulfuric, hydrochloric, carbonic, and boric acids. The inclusion of these classes evidences a trend in early nineteenth-century mineralogy to widen the scope of the science of mineralogy to encompass not only all chemical combinations, such as gases, but also rocks. It was much later that the term *mineral* was limited to designate a naturally occurring element or compound formed as a product of inorganic processes. The enumeration of air among non-acidic oxides reflects the early nineteenth-century view that air was a chemical combination of gases rather than a mixture. The inclusion of hydrochloric acid among acidic oxides evidences the contemporary conviction that all acids possessed chemically combined oxygen.

Brongniart's second departure from the system of Haüy occurred in his third class, termed "stones or earths combined with themselves and sometimes with accessory alkaline, acid, or metallic principles." Because Haüy had insisted that the so-called electropositive elements ought to be employed to designate genera throughout an entire mineral classification, he had been unable to make a sensible classification of the silicates and had resorted merely to listing various species. Brongniart overcame this difficulty by designating as genera combinations of elements and compounds commonly found in the various oxides and silicates, such as "silica, alumina, and alkali," "silica, alumina, calcium, and water," "silica, alumina, and beryllium," and so forth. Under each of these genera he included various species; for example, under those just named were listed lazurite, minerals of the zeolite family, and beryl, respectively.[35]

Brongniart's system of classification was adopted either wholly or on a modified basis by many mineralogists. Only very gradually, however, did mineralogists begin to realize that all properties of minerals had to be taken into account in establishing a system of classification: the crystal class, the external characters, and the proportionate relations between the bases and acids they contained, as well as

[34] R. J. Haüy, *Traité de minéralogie* (4 v., Paris, 1801).

[35] A. Brongniart, *Traité élémentaire de minéralogie* (2 v., Paris, 1807).

isomorphous substitutions. But this conception was not present in the early years of the nineteenth century. Mineral classification was in a state of flux, and many more systems of classification were to be introduced and discarded before the passion for meticulous taxonomy waned.

II. EUROPE AND AMERICA

By the end of the eighteenth century mineralogy was a well-established science replete with rival schools of thought, government and private patronage, and rapidly multiplying means of communication. Until the French Revolution, The *Philosophical Transactions* of the Royal Society of London and the *Mémoires* of the Académie Royale des Sciences were the two most important avenues for the communication of scientific discoveries. The *Journal de Physique* was a secondary publication to which scientists could turn and which often published extracts of papers that would later appear *in extenso* in the *Mémoires*. However, Jean Claude de la Métherie, the editor of the *Journal de Physique* from 1785 until his death in 1817, was a severe critic of Antoine Lavoisier's new oxygen chemistry and its novel nomenclature. As a consequence, Lavoisier and his colleagues, with the approval of the Académie des Sciences, established in 1789 the *Annales de chimie et de physique,* which rapidly became an influential and important scientific journal. After the reorganization which led to the formation of the Conseil des Mines in 1794 the *Journal des Mines* made its appearance, serving the community of geologists and mineralogists. During the Napoleonic period the *Annales du Muséum National d'Histoire Naturelle* also published many articles devoted to mineralogy.

In the British Isles, similarly, the rapid development of the sciences proved too much for the pages of the *Philosophical Transactions*. The *Philosophical Magazine* took over some of the burden in 1789, and other journals followed in the early years of the nineteenth century. The most important in the area of mineralogy were *Nicholson's Journal, Annals of Philosophy,* the *Edinburgh Philosophical Journal,* and the *Edinburgh Journal of Science.* In Germany, A. F. Gehlen's *Journal für die Chemie und Physik* appeared at the beginning of the nineteenth century. Karl Cäser von Leonhard established the *Taschenbuch für die gesammte Mineralogie* in 1807, which soon rivaled the *Journal des Mines* in importance. After the rejuvenation of the Berlin Academy, the *Abhandlungen der Königlichen Akademie der Wissenschaften in Berlin* began publication.

It was not unusual for these journals to translate or make extracts of articles that had been published in journals in other languages. A number of important books or doctoral dissertations written in Latin, such as Torbern Bergman's *Opuscula physica et chemica*

and C. S. Weiss's *Dissertatio de Indagando formarum cristallinarum caractere geometrico,* were translated into the vernacular. But many important works, such as Haüy's *Traité de minéralogie,* were never translated, so that scientists had to acquire a reading knowledge of at least two of the three languages of scientific discourse—English, French, and German.

On the methodological level, three main approaches to the study and classification of minerals were firmly established. In Germany, Werner's system of classification by external characters was predominant. In France, Haüy's crystallographic method reigned supreme. In Sweden, the chemical analysis of minerals, pioneered by Torbern Bergman, had taken firm root. For the next several decades these rival approaches to the study of minerals were to contend for supremacy until, in due time, the legitimate claims of each were recognized.

In the opening decades of the nineteenth century the chief source of controversy was the growing challenge of the new chemistry, inaugurated by Antoine Lavoisier and his colleagues in the 1780's, to Haüy's sweeping claims for his crystallographic method. This challenge was given added impetus by two important developments: the rapid discovery of new elements and the revival of the atomic theory. Thirty elements had been identified by 1800, seven of which had been known since ancient times. By 1828 another twenty-one had been discovered, although the isolation of several of these from their compounds had to await the advent of more sophisticated techniques. Several of the new elements were isolated by electrochemical methods. In 1807 and 1808, Sir Humphry Davy discovered sodium, potassium, calcium, and magnesium by placing their slightly damp fused oxides between the poles of a large storage battery. The majority, however, were brought to light as a result of the ever-increasing improvement of analytical techniques. With the discovery of these new elements mineralogists began to recognize that minerals, in particular the silicates, were much more complex chemically than had been thought and that classifying them by the methods of botanical taxonomy was a practical impossibility.

On the theoretical level, the major chemical advance of the early nineteenth century was John Dalton's theory that the atoms of different elements have different weights, and that chemical combination between elements occurs by the union of single atoms or a multiple number of them. Experimental evidence suggested that Dalton's ideas were sound. The problem was to determine the relative weights of the atoms and to find the number of simple atoms forming a compound. The leading figure in the attack on this problem was Jöns Jakob Berzelius, the great Swedish chemist whose laboratory in Stockholm was becoming a Mecca for students of chemistry. It was Berzelius who did most of the work in establishing relative atomic weights and ascertaining the chemical formulas of inorganic and organic compounds. A master of quantitative and qualitative analysis, Berzelius sought to base mineralogy squarely on chemical science. Through his own work and the discoveries of his students and disciples Berzelius presented a major challenge to Haüy's purely crystallographic approach to the classification of minerals.

One of the most important tools employed by Berzelius in his determination of atomic weights was the law of isomorphism. Isomorphism is the mutual substitution of two different chemical elements in a single chemical compound, the atomic sum of the two standing in the same quantitative relations to the other constituents as if one of the elements had been present alone. This phenomenon had been noted by Johann Fuchs in gehlenite in 1815, and had puzzled François Sulplice Beudant, one of Haüy's students, in 1817. Haüy had declared that mineral species should be defined by two types: the geometrical, consisting in the figure of the integrant molecule, and the chemical, grounded in the chemical composition of the integrant molecule. Of the two, the former should be the ultimate criterion of the species in case of doubt. In accordance with this position, Haüy stated that the integrant molecule of iron sulfate was an acute rhombohedron, that of copper sulfate an irregular oblique-angled parallelepipedon, and that of zinc sulfate a regular octahedron. Beudant, however, determined that solutions containing 85 per cent zinc sulfate and 15 per cent iron sulfate and those which contained 90 per cent copper sulfate and 10 per cent iron sulfate still yielded crystals having the rhombohedral figure of pure iron sulfate. Even solutions of 97 per cent mixed zinc and copper sulfate and 3 per cent iron sulfate produced rhombohedral crystals. Haüy's explanation of these phenomena was that in some chemical mixtures one constituent directed or had a dominant influence in the production of the crystals precipitated from the solution to the exclusion of the other constituents.[1]

In 1818 William Hyde Wollaston disputed Haüy's explanation, but it was Eilhard Mitscherlich, a coworker of Berzelius and later professor of chemistry at Berlin, who provided the solution to the problem in 1819 when he stated the law of isomorphism. Mitscherlich began his work with the study of phosphorous and arsenious salts, both of which combined oxygen in the same proportions and both of which had identical amounts of water of crystallization. He determined that the combinations of these salts with the same metallic bases resulted in the production of crystals having the same figure. Extending his research, he found that the oxides of magnesium, copper, and nickel, when combined with the same acid, crystallized in the same configuration. At his point Mitscherlich decided to repeat Beudant's experiments with the sulfates. He then discovered that the same crystal figure was produced when the same proportionate amount of water of crystallization was present. On the basis of these results Mitscherlich reported to the Berlin Academy of Sciences that any chemical combination in which the number of atoms was the same assumed the same crystal figure on crystallization. The crystal figure was independent of the chemical nature of the atoms. Mitscherlich's paper included the first definitive statement of the law of isomorphism, although he was aware that some modifications might be necessary after future research.[2]

Mitscherlich's later studies of crystal form provided the solution to another problem that had concerned chemists and mineralogists for more than two decades. This discussion focused on calcite and aragonite, which had completely dissimilar integrant molecules but apparently the same chemical composition. Because the two minerals did not have the same integrant molecules, nor ones that could be mathematically related, Haüy termed them different mineral species, whereas chemists, after hundreds of analyses which showed the two to be chemically identical, insisted that they should be classed as the same species. Haüy and his adherents, however, viewed the impasse as an outstanding proof that chemical analysis should sometimes be abandoned in favor of crystal form as a means of determining mineral species. It seemed that Haüy was the victor when Friedrich Stromeyer announced that aragonite contained a minute amount of strontium carbonate which had not been found in calcite. But Stromeyer's finding was unconvincing because analyses of other specimens of aragonite did not reveal the presence of strontium carbonate.[3]

Extending his studies of isomorphism, Mitscherlich determined that calcite, magnesite, and iron carbonate were isomorphous, as were aragonite, strontium carbonate, and lead carbonate. He then revised his law, stating that an identical number of atoms combined

[1] R. J. Haüy, *Traité de minéralogie* (5 v., 2nd ed., Paris, 1822) 1: pp. 25–26; François Sulplice Beudant, "Recherches tendantes à déterminer l'importance relative des formes cristallines et de la composition chimique dans la détermination des espèces minérales," *Annales de Chimie et de Physique* 4 (1817): pp. 72–84; Georges Cuvier, "Ânalyse des travaux de l'Académie Royales des Sciences pendant l'année 1817, partie physique," *Mémoires de l'Académie Royale des Sciences* 1817: p. ciii.

[2] Eilhard Mitscherlich, "Ueber die Kristallisation der Salze in denen des Metal der Basis mit zwei proportionen Sauerstoff verbunden ist," *Abhandlungen der Königlichen Akademie der Wissenschaften in Berlin* 1818–1819: pp. 427–437.
[3] R. J. Haüy, "Sur l'arragonite," *Annales du Muséum d'Histoire Naturelle* 11 (1808): pp. 241–270; "Discovery of the Composition of Arragonite," *Philos. Mag.* 42 (1813): p. 25.

in the same manner will produce the same crystal figure; the figure does not depend upon the nature of the atoms but rather upon their number and their manner of combination. Thus, calcite and aragonite were chemically identical, but their atoms were arranged in a different manner. Mitscherlich called the latter phenomenon dimorphism, later termed polymorphism.[4]

The discovery of isomorphism dealt a death blow to Haüy's definition of mineral species in terms of the shape of the integrant molecule. The figure of the integrant molecule could be almost exactly the same if the number of chemically related atoms and their manner of aggregation were identical in two or more substances. Similarly, polymorphism demonstrated that the same chemical species could exist in at least two different forms that were entirely unrelated by Haüy's law of decrement. The figure of the integrant molecule, then, was henceforth but one of the many characteristics useful in identifying minerals. Haüy, who by this time was nearing the end of his career, could not accept the results of Mitscherlich's work. If Mitscherlich's theory was correct, he confided to a friend, mineralogy would be the most wretched of the sciences. In a letter to Mitscherlich, Berzelius explained Haüy's reaction:

One ought not to expect that a gray-haired scientist close to the end of an honorable life should give up, without resistance or any attempt at justification, a theory he erroneously considered to be the most important of his discoveries; this is perhaps too much to demand morally of any man.[5]

Advances in optics also aided the theoretical development of the science of mineralogy. All crystals except those belonging to the isometric system, in which the crystals are described by reference to three mutually perpendicular axes of equal length, display the phenomenon of the double refraction of light. In crystals belonging to the tetragonal and hexagonal systems, a light ray traveling parallel to the vertical crystallographic axis encounters an isotropic structure, so that it will not be doubly refracted. Traversing the crystal in any other direction, however, it meets an anisotropic structure, which causes the light ray to split and pass through the crystal with two different velocities. Because there is only one unique direction of isotropy in crystals belonging to these two systems, the crystals are termed optically uniaxial. Similarly, symmetry causes crystals of the orthorhombic, monoclinic, and triclinic systems to have two axes along which double refraction does not

occur, and minerals belonging to these systems are called optically biaxial. In the majority of minerals, the existence of double refraction can be determined only by consciously searching for the phenomenon. In calcite, however, it is easily seen, a circumstance which led to the discovery of double refraction by Erasmus Bartholinus in 1669.

Although a number of scientists studied the phenomenon of double refraction after Bartholinus's memoir, understanding of the interconnection between optical properties and crystal form awaited the discovery of the polarization of light by Étienne Malus in 1809.[6] Shortly afterwards Jean Baptiste Biot and independently David Brewster discovered optical biaxiality and used it in the analysis of crystals.[7] Over a period of years Brewster classified about three hundred mineral crystals by whether they had one, two, or three axes of double refraction. He noted that all crystals with the cube, the regular octahedron, and the rhombic dodecahedron as a primitive form had three axes of double refraction; that is, such crystals showed no double refraction of light. Crystals having as primitive forms the rhombohedron, the hexagonal prism, or the octahedron with isosceles triangular faces had one axis of double refraction; substances with two axes of double refraction crystallized in other irregular forms. Since the phenomenon of double refraction delineated the position of the crystal axis, it could be used to determine the primitive form and that of the integrant molecule. Brewster believed that it was much more difficult to determine the primitive form by cleavage, as Haüy's method demanded, than by testing the crystal for double refraction. Soon Brewster began to publish lists of minerals which could not possess the primitive forms assigned to them by Haüy because their optical characteristics were not compatible with those forms.[8]

Further experimentation gave ample evidence that mineral crystals had unique axes in a physical sense. In 1824 Eilhard Mitscherlich described the effects of temperature change on the dimensions of a number of crystals, demonstrating that crystals that were regular geometric solids expanded uniformly in all directions upon an increase in temperature without showing any

[4] Eilhard Mitscherlich, "Ueber das Verhältnis zwischen der chemischer Zusammensetzung und das Kristallform arsenikersauerer und phosphorsauerer Salze," in: A. Mitscherlich, ed., *Gesammelte Schriften von Eilhard Mitscherlich* (Berlin, 1896), pp. 133–173.

[5] *Ibid.*, p. 173 fn.

[6] Étienne Malus, "Sur une propriété de la lumière réfléchie," *Mémoires de Physique et de Chimie de la Société d'Arcueil* 2 (1809) : pp. 143–158.

[7] Jean B. Biot, "Mémoire sur un nouveau genre d'oscillation que les molécules de la lumière éprouvent en traversant certains cristaux," *Mémoires de l'Institut* 1 (1812) : pp. 1–371; David Brewster, "On the Affections of Light Transmitted through Crystallized Bodies," *Philos. Trans. Royal Soc. London* 1 (1812) : pp. 187–218.

[8] David Brewster, "On the Laws of Polarization and Double Refraction in Regularly Crystallized Bodies," *ibid.* 1818: pp. 199–273; Brewster, "On the Connection Between the Optical Structure and the Chemical Composition of Minerals," *Edinburgh Philos. Jour.* 5 (1821) : pp. 1–8.

change in the value of their interfacial angles, while crystals whose primitive forms were rhombohedra or hexagonal prisms displayed unequal expansion when they were heated. Mitscherlich also showed that all crystals with a single axis of double refraction responded similarly to the action of heat and the action of light.[9]

But although chemistry and optics dissipated Haüy's dream of a purely crystallographic science of mineralogy, advances in crystallography continued to contribute powerfully to the progress of the science. The analysis of the external symmetry of crystals was taken up by Christian Samuel Weiss, who had studied mineralogy under Werner at Freiberg and had worked briefly with Haüy. Every crystal, Weiss declared, possessed a principal and dominating axis. Some crystals had more than one axis; the cube, for example, had three. The axes were not to be thought of as having merely geometrical existence; they should also be considered as physical entities. In Weiss's view, Haüy had erred in considering the relationship between the two diagonals of the face of the primitive form of calcite to be the basic element of the form. This relation was not fundamental because the principal geometric characteristic should be based on elements that had a direct relation with the crystallographic axis. The relation between the sines and cosines of the angles of inclination of a face to the axis did have a direct relation; hence the crystal form should be analyzed in terms of these direct relations.[10]

In December, 1815, Weiss read a memoir to the Berlin Academy in which many of the modern aspects of crystal morphology were presented for the first time.[11] He classified crystals into two main divisions: those that had three axes at right angles to one another and those that had four axes, three of which were equally separated and perpendicular to the fourth. In the first division, the three axes could have equal lengths; two might be equal and differ from the third; or the lengths of all three could differ. Thus, Weiss distinguished the present isometric, tetragonal, orthorhombic, and hexagonal crystal systems.

The principal forms of the isometric system were the cube, the regular octahedron, and the rhombic dodecahedron. Thus, Weiss recognized that three of Haüy's

six primitive forms possessed the same type of symmetry. The three equal mutually perpendicular axes created eight octants; based on symmetry considerations, the maximum possible number of similar faces at different inclinations was six. Weiss concluded, therefore, that the maximum number of possible similar plane surfaces in the isometric system was forty-eight. He saw, however, that further subdivision of this system was necessary. One class should include those forms in which all similarly shaped surfaces were presented, thus distinguishing the hexoctahedral or holohedral class. He correctly listed diamond, fluorite, spinel, and garnet as examples. When all of the similarly shaped faces were not present, the forms should be included in one of two subdivisions, corresponding to the hextetrahedral and diploidal classes of the isometric system. Weiss correctly identified sphalerite as an example of the former and pyrite as an example of the latter.

Weiss demonstrated that the crystal forms of the present tetragonal system could have a maximum of sixteen similar surfaces. But since he took only axial symmetry operations into account, he was unable to distinguish any subdivision of this or any other crystal system which introduced other symmetry operations. He did list zircon, zinc sulfate, and rutile correctly as members of this system.

Weiss divided the crystal system with three unequal mutually perpendicular axes into three classes, one of them corresponding to the orthorhombic crystal system. He recognized that crystals of this system might have eight similar surfaces and included topaz, niter, and aragonite as examples. But in retaining mutually perpendicular axes for the other two classes, which would otherwise correspond to the triclinic and monoclinic systems, he introduced a serious flaw. Nevertheless, he did correctly list chalcanthite as an example of the former, and feldspar, gadolinite, and borax as examples of the latter. Two years later, Weiss developed a new notational method for describing crystal forms. The modern Miller indices are the reciprocals of Weiss's symbols; they were introduced by the English crystallographer William Hallowes Miller in 1839.

Contemporaneously with Weiss, Friedrich Mohs, Werner's successor at Freiberg, was also studying the possibility of establishing crystal systems based on axial symmetry. In 1822 he published the first volume of his *Grund-Riss der Mineralogie*, in which he classified crystals into systems as Weiss had done, but he advanced one step further in recognizing the possibility of systems in which the axes were not all mutually perpendicular. Two years later, in the second volume, Mohs, and independently Karl F. Naumann, identified the modern monoclinic and triclinic systems. From that time on, the external symmetry elements of crystals were assiduously studied, and mineralogists began to identify the crystal system and

[9] Eilhard Mitscherlich, "Ueber das Verhältnis der Form der kristalliserten Körper zur Ausdehnung die Wärme," A. Mitscherlich, ed., *Gesammelte Schriften von Eilhard Mitscherlich*, p. 195.

[10] Christian Samuel Weiss, "Dissertatio de Indagando formarum cristallinarum caractere geometrico, ou Mémoire sur la détermination ou caractère géometrique principal des formes cristallines," Brochant de Villiers, tr., *Journal des Mines* 19 (1811): pp. 349–391, 401–444.

[11] C. S. Weiss, "Uebersichtliche Darstellung des verschiedenen natürlichen Abteilungen der Kristallisationssysteme," *Abhandlungen der Königlichen Akademie der Wissenschaften in Berlin* 1814–1815: pp. 289–344.

the subdivision of the system to which a mineral belonged as a part of the description of the mineral.[12]

The existence of only fourteen space lattices was not definitely proved until 1848, when Auguste Bravais deduced them mathematically. By the 1830's, however, a number of mineralogists were seeking to explain the external symmetry of mineral crystals by the internal arrangement of atoms, molecules, or point forces. Dalton in 1808, Wollaston in 1813, and John F. Daniell in 1816 and 1817 all attempted to reestablish the view originally proposed by Hooke and Huygens in the seventeenth century, that the ultimate particles of matter were spheres and that the various modes of stacking these spheres could account for the diverse configurations of crystals.[18] In Germany, Johann Prechtl objected to Haüy's integrant molecules on the ground that the surfaces of the tiny polyhedra would necessarily reflect incident light. Instead he postulated that crystallization was produced by the reciprocal attraction of homogeneous and similar globules formed from the supernatant fluid.[14]

A further attack on Haüy's molecular theory came from Ludwig Seeber in 1824. Polyhedral forms were unnatural and improbable, Seeber said. The hypothesis of polyhedral forms was a consequence of Haüy's explanation of the natural cleavage faces of crystals and his idea that these faces were contiguous. Since Haüy's concept encountered the greatest difficulties when the integrant molecules were not parellelepipedons, Seeber proposed to substitute minute spherical atoms in the centers of Haüy's integrant molecules. These would in effect trace a space lattice.[15] Meanwhile two French mathematicians, Claude Navier and Augustin Cauchy, investigating problems of elasticity, suggested the idea of conceiving solids as arrays of mathematical points extended symmetrically throughout space. In 1848 these researches, as well as those of Moritz Frankenheim and G. Delafosse, were brought to fruition by Auguste Bravais in his work on space lattices.[16] Crystallography and atomic theory

were now joined in a partnership that was to prove of great significance, not only for mineralogy but for physics and chemistry generally.

Viewed against the backdrop of these impressive developments in European science, the achievements of mineralogical science in the United States in the same period pale into relative insignificance. Before the American mineralogists lay a vast continent whose mineral resources invited discovery, description, and exploitation. But, despite confident assertions that republican institutions would prove far more favorable to the development of science and industry than the monarchical and aristocratic institutions of Europe, the hard fact was that American society had no ready replacement for the European system of patronage by government subsidy and the private munificence of a leisure class. The British models on which the American governments were formed afforded no precedent for extensive governmental support of science and industry, and the Jeffersonian principles which soon prevailed in American politics reduced still further the prospect of public patronage for either. As for private benefactions, the conditions of life in America militated against the early development of a wealthy leisure class willing and able to support scientific investigation. Everything—scientific societies, libraries, collections, colleges, and universities—had to be built from scant beginnings. In these circumstances it is not surprising that the science of mineralogy acquired a solid footing in the United States only gradually, remaining on the descriptive level well into the nineteenth century.

Some idea of the extent of American familiarity with the progress of mineralogy in Europe and of the obstacles confronting American aspirations in this field of inquiry may be gained from a book published in New York in 1803 entitled *A Brief Retrospect of the Eighteenth Century*. The author, a young Presbyterian clergyman of broad and solid learning named Samuel Miller, had undertaken to provide, "a sketch of the revolutions and improvements in science, arts, and literature" during the century that had just expired. In his discussion of mineralogy, which he classified as a branch of natural history, he showed himself familiar with the three basic approaches to the study of minerals described above. To Abraham Gottlob Werner he gave credit for reviving the description and classification of minerals in terms of their external characters and for supplying a nomenclature "more precise, accurate, and scientific than mineralogists had ever before possessed." For the development of the chemical analysis of minerals he had special

[12] Friedrich Mohs, *Grund-Riss der Mineralogie* (2 v., Dresden, 1822-1824) 1: p. 56 ff.; 2: pp. v–ix; K. F. Naumann, *Okens Isis* 9: pp. 954–959.

[13] John Dalton, *A New System of Chemical Philosophy* (2 v., Manchester, 1808) 1: p. 211; William Hyde Wollaston, "On the Elementary Particles of Certain Crystals," *Philos. Trans. Royal Soc. London* 1813: pp. 51–63; J. F. Daniell, "On Some Phenomena Attending the Process of Solution and Their Application to the Laws of Crystallization," *Quart. Jour. Science* 1 (1816): pp. 24–49; Daniell, "On the Theory of Spherical Atoms and on the Relation Which It Bears to the Specific Gravity of Certain Minerals," *ibid.* 4 (1817): p. 30.

[14] Johann Prechtl, "Théorie de la cristallisation," *Journal des Mines* 28 (1810): pp. 261–312.

[15] Ludwig Seeber, "Versuch einer Erklärung des inneren Baues der festen Körper," *Gilberts Annalen der Physik* 76 (1824): pp. 229–248, 349–372.

[16] Auguste Bravais, "Les systèmes formés par des pointes distribués régulièrement sur un plan ou dans l'espace," *Journal de l'École Polytechnique* 19 (1850): pp. 1–128; Moritz Frank-

enheim, *Die Lehre von der Cohäsion* (Breslau, 1835); G. Delafosse, "Recherches relatives à la cristallisation, considérée sous les rapports physiques et mathématiques," *Académie des Sciences, Comptes Rendus* 11 (1840): pp. 394–400.

praise for the work of Axel Fredrik Cronstedt, superintendent of mines in Sweden, for Cronstedt's countrymen Carl Scheele and Torbern Bergman, and for
Martin Klaproth in Berlin and Louis Nicolas Vauquelin in Paris. The new crystallographic method of
analyzing minerals he attributed correctly to Romé
de l'Isle and the Abbé Haüy, acknowledging their
work to be "a monument of ingenuity and labour . . .
supposed by some to give important light, and to
promise much usefulness, both in mineralogy and
chemistry." He concluded, however, that most of the
leading mineralogists based their nomenclature and
classification on a combination of external characters
and chemical properties. Among these he gave special
praise to Richard Kirwan of Ireland, whose works
were well known in the United States and whose collection of minerals, assembled originally by Professor
Nathaniel Gottfried Leske of Leipzig, Miller considered "the best collection on earth." [17]

Of the practical importance of mineralogical science
for national economies and industries Miller was fully
convinced. Mineralogy, he declared,

has lately begun to be viewed as dignified in its nature,
and most interesting in its relations. It is now regarded
as a valuable and indeed necessary handmaid to *Medicine,
Agriculture,* and a large portion of the *manufactures,*
which supply the conveniences, comforts, or luxuries of
human life. Mineralogy has therefore, within a few years
past, been cultivated with great diligence and success by
almost all the nations in Europe, especially in *Germany*
and *Sweden,* where splendid mineral riches particularly
invited inquiry and application. Societies have been
formed for extending and improving the science; travel-
lers have explored foreign countries for the same purpose; distinguished eminence in this branch of knowledge
has been rewarded by public esteem, and by civil honours;
and the most effectual methods used to direct general
attention to the subject.[18]

In the United States, Miller conceded, mineralogical
science was in its infancy despite the vast mineral
treasures waiting to be discovered, described, classified, and utilized. In this field of investigation, as in
all science, literature, and art, Americans were handi-
capped by the newness of their country and its institutions. Of scientific societies Miller could list only
two, the American Philosophical Society in Philadelphia and the American Academy of Arts and Sciences
in Boston. Among the colleges, none was handsomely
endowed compared with European universities, and
specialized research was difficult or impossible. In
society generally, a wealthy leisure class willing and
able to devote itself to science, literature, and the arts
was lacking.

The comparatively equal distribution of property in
America, while it produces the most benign political and
moral effects, is by no means friendly to great acquisitions
in literature and science. In such a state of Society, there
can be few persons of leisure. It is necessary that almost
all should be engaged in some active pursuit. Accord-
ingly, in the United States, the greater number of those
who pass through a course of what is called liberal
education . . . engage, immediately after leaving College,
in the study or business to which they propose to devote
themselves. . . . Such is the career of ninety-nine out of
a hundred of those in our country who belong to the
learned professions. When the alternative either lies,
or is supposed to lie, between erudition and poverty, or
comfortable affluence and moderate learning, it is not
difficult to conjecture which side will be chosen. . . .

To this circumstance may be ascribed the superficial
and unpolished character of many of our native publications. All that their authors, in many cases, want, to
render them more replete with instruction, more attractive
in manner, and, of course, more worthy of public approbation, is *leisure.* But, able only to redeem a few hasty hours
for literary pursuits, from the employments which give
them bread, they must necessarily, if they publish at all,
send forth productions, from time to time, bearing all the
marks of haste and immature reflection.[19]

To these disadvantages, Miller continued, were
added a dearth of books, libraries, and scientific collections, a commercial spirit unfavorable to abstruse
studies, and a lack of inducements to their pursuit.
But times and conditions were changing, and the outlook for the future improving rapidly.

The emulation of founding and sustaining a national
character in science and learning begins to be more
generally felt. . . . A larger proportion of the growing
wealth of our country will hereafter be devoted to the
improvements of knowledge, and especially to the furtherance of all the means by which scientific discoveries are
brought within popular reach, and rendered subservient
to practical utility. American publications are every day
growing more numerous, and rising in respectability of
character. . . . American authors of merit meet with more
liberal encouragement, and when the time shall arrive
that we can give to our votaries of literature the same
leisure, and the same stimulants to exertion with which
they are favoured in Europe, it may be confidently pre-
dicted, that letters will flourish as much in America as
in any part of the world; and that we shall be able to
make some return to our transatlantic brethren, for the
rich stores of useful knowledge which they have been
pouring upon us for nearly two centuries.[20]

Miller's assessment of the status and prospects of
American science and letters was well founded. The
wheel horses of American science were indeed men
engaged in practical pursuits that left them relatively
little time for scientific research. Physician-scientists
like Benjamin Smith Barton, Samuel Latham Mitchill,
and Benjamin Waterhouse practiced medicine, taught
medical students and liberal arts undergraduates, often
undertook to edit medical journals, and, in the case of
Mitchill, served in the state legislature and in Con-

---

[17] Samuel Miller, *A Brief Retrospect of the Eighteenth Century, Part First In Two Volumes Containing A Sketch of the
Revolutions and Improvements In Science, Arts, and Literature During That Period* (2 v., Facsimile edition, Burt Franklin, New York, 1970; originally 1803) 1: pp. 145–155.
[18] *Ibid.* 1: p. 155.

[19] *Ibid.* 2: pp. 405–406.
[20] *Ibid.* 2: pp. 409–410.

gress. Professors in the liberal arts colleges, like Benjamin Silliman and Parker Cleaveland, taught a broad range of subjects in addition to performing administrative functions. Clergyman-scientists like Henry Muhlenberg and Manasseh Cutler combined their scientific researches with parish duties. Cutler also served in Congress. The astronomer David Rittenhouse was a surveyor and treasurer of the State of Pennsylvania; Nathaniel Bowditch was a mariner before he made his reputation as a practical astronomer and went into the insurance business.

But these wheel horses of American science received substantial assistance from two other kinds of scientists. On the one hand, there were a number of persons who might be called gentlemen scientists—men of substantial means who decided to devote themselves to scientific research for its own sake. Among these were Stephen Elliott in Charleston; Zacchaeus Collins, Solomon W. Conrad, Charles Wister, and Adam Seybert in Philadelphia; Robert Gilmor, Jr. in Baltimore; Benjamin Vaughan in Hallowell, Maine; and Colonel George Gibbs in Newport, Rhode Island. On the other hand, there were a considerable number of foreign scientists who emigrated to the United States or came here temporarily to explore the natural history of North America and win scientific glory by describing the minerals, plants, and animals of the New World. Some, like William Maclure, were men of substantial fortune, but most, like Thomas Nuttall, Alexander Wilson, Dr. Thomas Meade, and C. S. Rafinesque, were scientific adventurers spurred on by dreams of scientific reputation and a keen zest for the life of a roving naturalist. In the end, all of these types of researchers were to contribute handsomely to the development of American mineralogy.

Finally, it should be made clear that American science, unlike that in European countries, developed regionally. With the removal of the national capital to Washington, D. C. in 1800, all hope vanished of developing a capital city which, like London or Paris, would be the scientific, artistic, and literary center of the nation. Instead, each region developed its own cultural center or centers—Charleston in the South, Philadelphia, New York and Albany in the middle states, New Haven, Cambridge, and Boston in New England. As a result, the story of the growth of mineralogy in the United States must be told regionally, beginning with the oldest and most vigorous scientific center in Philadelphia, proceeding thence to New York and Albany and onward to Boston and Cambridge, and concluding with the emergence of mineralogy and geology as subjects of study in the liberal arts colleges of Bowdoin and Yale. Eventually, however, all parts of the nation from Savannah to Maine and as far west as Lexington and Nashville were to make contributions to the rise of mineralogy in the American republic.

III. MINERALOGY IN PHILADELPHIA

Among the early scientific centers in the United States, Philadelphia was pre-eminent.[1] It was, as John Adams remarked somewhat enviously, "the pineal gland of the republic." Besides the American Philosophical Society, the oldest scientific society in the former British colonies, it boasted Peale's museum of natural history, the Pennsylvania Hospital, a chemical society, several gardens of botanical interest,[2] a Linnaean Society (later superseded by the Academy of Natural Sciences), and a university with an outstanding medical school. Of these institutions the one of immediate importance for the development of American mineralogy was the School of Medicine of the University of Pennsylvania, for it was in medical schools that mineralogy first obtained academic status in Britain and America as an adjunct of the study of chemistry and natural history. As a science which aimed at naming, classifying, and describing minerals according to their external characters, mineralogy was deemed a branch of natural history. Insofar as it sought to base its nomenclature and classification on chemical analysis and crystallographic structure, however, it allied itself with natural philosophy, but not with those branches of natural philosophy that had long formed part of the curriculum of the liberal arts colleges. Thus mineralogy, like its foster parents, chemistry and natural history, had to find its way into the liberal arts course through the medical schools.

Not surprisingly, the first devotees of mineralogical science in Philadelphia were men trained in chemistry at the medical school in that city. Four men in particular—James Woodhouse, John Redman Coxe, Adam Seybert, and Robert Hare, Jr.—had intertwining careers connected with the chemical chair at

---

[1] The present chapter draws liberally on John C. Greene, "The Development of Mineralogy in Philadelphia, 1780–1820," *Proc. Amer. Philos. Soc.* 103 (August, 1969): pp. 283–295.

[2] The Philadelphia naturalist John Bartram, who founded a botanical garden in 1730, was a collector of rocks and minerals. When Johan David Schoepf visited the garden in 1783, Bartram's son showed him "all manner of rocks and minerals which are now kept in a box without any system intermixed with European specimens, especially Swedish, sent over by Linnaeus Archiater." (See Schoepf's *Travels in the Confederation* [1783–1784] . . . Alfred J. Morrison, tr. and ed. (2 v., Philadelphia, 1911) 1: p. 92.) Another early Philadelphia collector was the merchant Samuel Hazard, who visited western Pennsylvania in connection with a project to colonize and Christianize the Indians. Benjamin Franklin thought Hazard's "large collection of Ores, Minerals, & other Fossils of these Parts of America" of sufficient importance to send a catalog of the items in it to England with the information that the collection was deposited in the Library Company of Philadelphia. See Benjamin Franklin's letter to Richard Jackson, Philadelphia, March 8, 1764, as quoted in Carl van Doren, ed., *Letters and Papers of Benjamin Franklin and Richard Jackson 1753–1785,* Memoirs Amer. Philos. Soc. 24 (1947): pp. 142–143. There is a brief biography of Samuel Hazard in *Appleton's Cyclopedia of American Biography* (7 v., New York, 1888–1900) 3: p. 149.

the School of Medicine and the early development of mineralogy in Philadelphia. Woodhouse took his M.D. in 1792, Seybert in 1793, and Coxe in 1794. Robert Hare studied chemistry under Woodhouse but did not take a medical degree. Woodhouse and Coxe were *protégés* of the prestigious and powerful Dr. Benjamin Rush, Seybert of the somewhat milder Dr. Caspar Wistar. When the chemical chair fell vacant in 1794 and was declined by Joseph Priestley, who had recently arrived in America, Rush nominated Woodhouse in preference to Coxe, who was then in London, and secured his election over the rival candidates Seybert and Benjamin Smith Barton. Seybert soon left for Europe to continue his studies at Edinburgh, London, Paris, and Göttingen, and Woodhouse, elated by the honor conferred upon him, threw himself into his new duties with feverish zest.

One of the medical students in Philadelphia at that time, Charles Caldwell, has left an unforgettable picture of Woodhouse at work in his laboratory in Anatomical Hall, a small University building on South Fifth Street opposite the State House yard:

... his devotion to chemistry and the labor he sustained in the cultivation of it were perfectly marvelous—not to say preternatural [Caldwell recalled in his *Autobiography*]. . . . During an entire summer (one of the hottest I have ever experienced), he literally lived in his laboratory, and clung to his experiments with an enthusiasm and persistency which at length threw him into a paroxysm of mental derangement, marked by the most extravagant hallucinations and fancies. . . .

The special object of his experiments at that time was the decomposition and recomposition of water. . . .

As already mentioned, the weather was almost unprecedentedly hot; and his laboratory was in sundry places perpetually glowing with blazing charcoal, and red-hot furnaces, crucibles, and gun-barrels, and often bathed in every portion of it with the steam of boiling water. Rarely, during the day, was the temperature of its atmosphere lower than 110° to 115° of Fahrenheit—at times, perhaps, even higher.

Almost daily did I visit the professor in that salamander's home, and uniformly found him in the same condition—stripped to his shirt and summer pantaloons, his collar unbuttoned, his sleeves rolled up above his elbows, the sweat streaming copiously down his face and person, and his whole vesture drippingly wet with the same fluid. He himself, moreover, being always engaged in either actually performing or in closely watching and superintending his processes, was stationed for the most part in or near to one of the hottest spots in his laboratory.[3]

A rather different picture of Woodhouse is provided by Benjamin Silliman, who was a fellow student with Robert Hare at Woodhouse's lectures on chemistry during the winters of 1802–1803 and 1803–1804.

In his person [wrote Silliman] he was short, with a florid face. He was always dressed with care; generally

he wore a blue broadcloth coat with metal buttons; his hair was powdered, and his appearance was gentlemanly.

His apparatus, Silliman adds, was humble but adequate. The classroom demonstrations, performed without assistance, were not always successful, however. "Our Professor had not the gift of a lucid mind, nor of high reasoning powers, nor a fluent diction; still, we could understand him, and I soon began to interpret phenomena for myself and to anticipate the explanations."[4]

Although Woodhouse was not primarily interested in mineralogical chemistry, he gave detailed instructions for experiments on iron and other metals in his *Young Chemist's Pocket Companion; Connected With a Portable Laboratory* (1797) and undertook a considerable number of mineral analyses himself. An extended example of his methods of analysis is provided in his response in the *Medical Repository* for 1798 to a letter from the Reverend James Hall describing what Hall supposed to be an artificial wall discovered under the surface of the earth in Iredell County, North Carolina. Some specimens of rock from this "wall" were brought to Philadelphia by Hall, and Woodhouse was able to subject them to a thoroughgoing chemical analysis. The stones were prismatic in shape and covered with a "brown ochreous matter" and a "fine white friable matter" similar to the incrustations found on basaltic rocks by European writers. After describing the general appearance of the stones, Woodhouse detailed the methods by which he arrived at the following analysis of their chemical content:

	Silex	Alumine	Lime	Magnesium	Iron	Loss
White friable powder	55	16	5	3	12	9
Brown ochreous crust	54	15	6	3	11	11
Powdered stone	58	16	6	3	10	7

Since these results agreed roughly with the analyses of basaltic rocks given by European chemists and since the external characters of the stones were likewise conformable to those of European basalts, Woodhouse concluded that the supposed "wall" was a natural basaltic formation. When this conclusion was questioned by the Reverend Zechariah Lewis of Connecticut, Woodhouse replied in full, citing a wide variety of European publications in support of his

---

[3] Charles Caldwell, *Autobiography of Charles Caldwell, M. D. With a Preface, Notes, and Appendix, by Harriot W. Warner* (Philadelphia, 1855), p. 175.

[4] George P. Fisher, *Life of Benjamin Silliman, M. D., LL. D. . . . Chiefly from his Manuscript Reminiscences, Diaries, and Correspondence* (2 v., New York, 1866) 1: p. 101. For more favorable accounts of Woodhouse's abilities as a teacher and lecturer see Edgar Fahs Smith, *James Woodhouse, A Pioneer in Chemistry 1770–1809* (Philadelphia, 1918), pp. 63–64. Silliman seems to have been put off by Woodhouse's "jocose" manner with the students and by his failure to illustrate the attributes of the Creator in his chemical lectures.

argument and concluding:

A mineralogist can distinguish basaltes [*sic*] from any other stone merely by inspecting them, as well as any other person can tell an apple from an orange, a pear from a peach. . . . The chemical analysis, likewise, of the solid stones, and what is improperly called rust and cement, demonstrates in a satisfactory manner, to chemists, that the stones are basaltic and there can be no appeal from the experiments, except by showing that they have been made in an improper manner.[5]

Woodhouse also published analyses of titanium ore, pyrolusite, sphalerite, zinc ore, and anthracite coal. According to his biographer, Edgar Fahs Smith, Woodhouse's interest in mineralogical analysis was incidental to his general interest in chemistry and derived in considerable part from his correspondence with Dr. Samuel Latham Mitchill, editor of the *Medical Repository,* and from association with Adam Seybert. It was Mitchill who sent him the large specimen of titanium ore which formed the subject of his article "Analysis of an Iron Ore, Containing Titanium" in Coxe's *Philadelphia Medical Museum* in 1808. Although Mitchill's colleague Dr. Archibald Bruce and his European correspondents were convinced that the specimen was chiefly the oxide of titanium combined with menachanite [ilmenite], Woodhouse found it to be composed of iron, titanium, lime, and alumine, without any silicious earth. "This analysis," wrote the twentieth-century chemist Edgar Fahs Smith, "is a real curiosity. At present it would not pass muster. Those who have analyzed titaniferous ores will fail to perceive in it the course they would pursue in arriving at the composition of this well-known substance."[6]

Woodhouse was equally unfortunate in his analysis of a specimen of pyrolusite sent to him by a Mr. Butland of Philadelphia. After analyzing the substance, Woodhouse concluded that it was "manganese of the first quality, containing little extraneous matter." His purpose in publishing this analysis, he declared, was "to induce gentlemen residing in the country, to pay some attention to the mineral productions of their fields, by which means they may greatly profit themselves and render the most important services to the arts, yet in their infancy in this part of the world."

Any person desirous of information, concerning any of our native fossils, by applying to me, shall be gratified, as far as it is in my power; and if the mineral sent to me is thought to be of any use to society, an accurate analysis of it shall be made free of expense.[7]

The same desire to promote the economic development of the new American nation ran through Woodhouse's memoir on anthracite coal and his account of the zinc mine on Perkiomen Creek about twenty miles from Philadelphia. The memoir on anthracite demonstrated by reference to numerous experiments the superiority of Lehigh anthracite over bituminous coal from Virginia and lauded chemical analysis as the means by which "important services may be rendered to our citizens, the arts benefited, and a foundation laid for a system of American mineralogy."[8] This memoir was widely quoted and acclaimed. The article on the Perkiomen zinc mine was less successful, however. Here again Woodhouse's chemical analysis of the ore (sphalerite) left much to be desired, and his conclusion that it could not be worked to commercial advantage provoked an acidulous reply from Adam Seybert, who by this time had returned from his mineralogical studies abroad and had begun to put his chemical knowledge to commercial use as a "druggist, chymist and apothecary" and a manufacturer of mercurials.[9]

Woodhouse also analyzed fragments of a meteorite which fell in Weston, Connecticut, early on the morning of December 14, 1807. Apparently Woodhouse wrote to a Connecticut acquaintance named Judson immediately upon learning about the meteor, urging him to send specimens of the meteorites to Philadelphia for chemical analysis and advertising himself as the only person who could perform this important scientific task satisfactorily. This letter seems to have given offense to Woodhouse's former pupil Benjamin Silliman, who was busy preparing his own account of the Weston meteor, for Woodhouse wrote Silliman an apologetic note on January 6, 1808, explaining that he had not known of Silliman's interest in the subject at the time he wrote Judson. He had received one of

[5] James Hall, "An Account of a Supposed Artificial Wall, Discovered under the Earth in North Carolina, in a Letter from the Rev. James Hall, A. M. to James Woodhouse, M.D. . . . ," *Medical Repository* 2 (1798): pp. 272–274. Woodhouse's reply follows pp. 275–278. Rev. Zechariah Lewis, "Remarks on a Subterranean Wall in North-Carolina . . . ," *Medical Repository* 4 (1801): pp. 227–234, followed by Woodhouse's reply and Lewis's response to that reply, *ibid.* 5 (1802): pp. 21–24, 397–407. Professor Clifford Frondel, Harvard University, notes: "I am unable to suggest the precise nature of the material. In any case the analyses are indeed close to an average basalt."

[6] Smith, *James Woodhouse*, pp. 233–234. James Woodhouse, "Analysis of An Iron Ore, Containing Titanium," *Philadelphia Medical Museum* 4, 3 (1808): pp. 206–207.

[7] James Woodhouse, "To the Editor of the Philadelphia Medical Museum," *Philadelphia Medical Museum* 1, 4 (1805): p. 450.

[8] James Woodhouse, "Experiments and Observations on the Lehigh Coal," *Philadelphia Medical Museum* 1, 4 (1805): p. 441.

[9] James Woodhouse, "An Account of the Perkiomen Zinc Mine, with an Analysis of the Ore," *Philadelphia Medical Museum* 5, 2 (1808): pp. 133–136; Adam Seybert, "Facts to Prove That . . . Sulphuret of Zinc May Be Worked With Advantage in the United States," *Ibid.* 5, 4 (1808): pp. 209–216; James Woodhouse, "Woodhouse's Reply to Seybert's Strictures . . . ," *ibid.* 6, 1 (1809): pp. 44–54. For a discussion of Woodhouse's analysis of the zinc ore, see Smith, *James Woodhouse*, p. 241 ff.

the meteorites, he told Silliman, and had undertaken its analysis:

I have examined the parts of the stone, attracted by the magnet, & the unattractable [sic] portion.
The attractable parts, appear to consist of iron, sulphur, & nickel. One hundred parts of the unattractable [sic] portion appear to consist of, about 50 parts of silex, 25 of iron, & from 4 to 6 parts of sulphur.

Nicle [sic] is undoubtably [sic] one of the component parts, as the solution of that part attracted by the magnet, gives to the muriatic acid, a green color.
It is almost impossible to determine the exact quantity of the magnesia. I think no lime is present.
The specific gravity of the specimen, I examined, was 3.696.
No malleable iron was present in the specimen which I examined, and no nitrous air rose from it, when dissolved in the nitric acid.
When you arrive in Philadelphia, I shall have more conversation with you, on this subject.[10]

Soon afterward Silliman arrived in Philadelphia with a manuscript account of the meteor written with his Yale colleague Professor James Kingsley, seeking an effective means of publishing this extensive memoir to the scientific world. Their more popular account, published in the *Connecticut Herald* and republished elsewhere, had already excited great interest. Apparently Silliman had forgiven Woodhouse's remarks in his letter to Judson, for he wrote Kingsley from Philadelphia on January 23:

I attended Woodhouse's lecture the day after I arrived. He received me politely, but made no allusion to the offensive part of his letter. He showed me his laboratory, which is a very fine one indeed. I dined with him yesterday and met a large party of *savans*. I cannot stay to relate many particulars. (Monday 25.) The meteor is immediately brought forward in every circle where I go. It was so at Woodhouse's. He was very modest, and even ridiculed the lunar theory [of the origin of meteors] which he advocated in his letter.[11]

Silliman knew that Woodhouse was preparing to publish a chemical analysis of the meteorites, but he seems to have decided not to show his own account to Woodhouse, preferring to consult instead with Woodhouse's old rival Adam Seybert:

Dr. Seybert is the only man here who has perused it. He gives it full credit and says that my views of the effects of heat, connected with the experiments, are demonstrative. I have not communicated the piece to ——, nor said anything about it. I am convinced, from what he has told me, that his own analysis was altogether loose and not to be depended on, nor am I at all afraid of any publication of his. Seybert advised me not to trust him; said he would play me some trick,—for instance, purloin and publish it as his own; and averred

that he did not know how to analyze a stone, and that he had not a single sure test or agent of any kind to do it with.[12]

Silliman's account was read before the American Philosophical Society on March 4, 1808, and referred to a committee of referees consisting of Woodhouse, Hare, and Joseph Cloud. On March 18 the committee recommended publication, and the memoir appeared in the sixth volume of the Society's *Transactions*, published in 1809. Meanwhile Woodhouse had published his own much briefer account in the *Philadelphia Medical Museum*. His description of the circumstances surrounding the fall of the meteorites was borrowed from the Silliman-Kingsley account, to which he alluded in his final paragraph. The results of his analysis were similar to those announced in his letter to Silliman but differed in several respects from Silliman's findings: [13]

	Woodhouse	Silliman
Silex	50	51.5
Iron ("attractable brown oxide of iron"—Silliman)	27	38
Magnesia	10	13
Sulphur	7	1
Nickel ("Oxide of nickel"— Silliman)	1	1.5
	95	105
	Loss 5	Gain 5
	100	

It is interesting to compare these findings with the results of a much more recent analysis of fragments of the Weston meteor by two modern chemists, Brian Mason and H. B. Wiik. From their investigation, published in 1965, it appears that the minerals contained in the meteorite are "olivine, orthopyroxene, plagioclase, nickel-iron, and chromite," with a phosphate mineral (apatite or merrillite, or both) present

[10] James Woodhouse to Benjamin Silliman, Philadelphia, January 6, 1808, Beinecke Library, Yale University.
[11] Benjamin Silliman to James Kingsley, Philadelphia, January 23, 1808, quoted in Fisher, *Life of Benjamin Silliman* 1: p. 224.

[12] *Ibid.*, p. 226. It is, of course, not absolutely certain that the ———— referred to in Silliman's letter is Woodhouse, but the circumstances surrounding the letter point strongly to that conclusion. Seybert's evaluation of Woodhouse must itself be evaluated in the light of the bad feeling between Woodhouse and Seybert, which was soon to be exacerbated by Seybert's acrimonious attack on Woodhouse's account of the Perkiomen zinc mine in the *Philadelphia Medical Museum* 5 (1808) : pp. 209–216; 6 (1809) : pp. 44–54. It should be noted, however, that Silliman adds in his letter to Kingsley: "————'s reputation is *up*, both here and at New York, for unfair dealing and in matters affecting scientific reputation."
[13] James Woodhouse, "Account of the Meteor Which Was Seen at Weston, in the State of Connecticut . . . ," *Philadelphia Medical Museum* 5, 2 (1808) : p. 132; Benjamin Silliman and James L. Kingsley, "Memoir on the Origin and Composition of the Meteoric Stones Which Fell from the Atmosphere, in the County of Fairfield, State of Connecticut, on the 14th of December 1807 . . . ," *Trans. Amer. Philos. Soc.*, O.S., 6, 2 (1809) : p. 339. See also "Early Proceedings of the American Philosophical Society . . . 1744 to 1838," *Proc. Amer. Philos. Soc.* 22 : 119 (1884) : pp. 404–405.

in accessory amounts.[14] Their chemical analysis, expressed in the conventional form as metal, troilite, and oxides, is as follows:

Fe	14.87	MgO	22.84
Ni	1.54	CaO	1.78
Co	0.11	Na$_2$O	0.95
FeS	5.31	K$_2$O	0.14
SiO$_2$	36.59	P$_2$O$_5$	0.32
TiO$_2$	0.13	H$_2$O+	0.98
Al$_2$O$_3$	2.23	H$_2$O−	0.10
FeO	11.14	Cr$_2$O$_3$	0.52
MnO	0.32		———
			99.87

Mason and Wiik conclude that the Weston meteorite is "an olivine-bronzite chondrite in Prior's (1920) classification." Thus, by modern standards the analyses of both Woodhouse and Silliman are erroneous. They did determine the principal chemical constituents of the meteorite, however, using methods and equipment vastly inferior to those available today. The extraordinarily high SiO$_2$ content reported by both Silliman and Woodhouse was undoubtedly the result of inadequate chemical separation from other elements, in particular aluminum.

Besides contributing mineralogical memoirs to the *Medical Repository* and the *Philadelphia Medical Museum*, Woodhouse founded and energized the Chemical Society of Philadelphia, which made the analysis of mineral specimens an important part of its business. Organized in 1792, the year in which Woodhouse received his M.D., the Society drew its membership primarily from the faculty, graduates, and students of the School of Medicine, although it also included some practical chemists like Robert Hare. Meeting weekly in Woodhouse's laboratory at Anatomical Hall, the Society issued a circular in 1798 inviting Americans throughout the nation to submit mineral specimens for analysis free of charge by a committee consisting of Woodhouse, Seybert, Thomas P. Smith, Samuel Cooper, and John C. Otto. In the *Medical Repository* for 1800 the committee reported that it had received specimens from many quarters, including a specimen of "golden auriferous pyrites" from Virginia, and added:

It is hoped that the importance of mineral substances in agriculture and manufacturing will induce the farmers and other gentlemen in the United States to attend to the mineral products of their fields and send them to the Chemical Society of Philadelphia, where they will be accurately analyzed, free of expense. By this means many valuable discoveries may be made and we may become

acquainted with the operations of nature in this part of the globe.[15]

The most important contribution to mineralogical science of the Chemical Society of Philadelphia was Robert Hare's invention of the oxyhydrogen blowpipe, an account of which was published by the Society in 1802, and reprinted the same year in Tilloch's *Philosophical Magazine* in London and in the *Annales de Chimie*.[16] The son of a well-known Philadelphia brewer, Hare was fifteen to twenty years younger than Woodhouse, Seybert, and Coxe. His formal schooling appears to have been limited, but he attended Woodhouse's chemical lectures and become a member of the Chemical Society of Philadelphia at an early age.

His enthusiasm for chemical experimentation rivaled that of his teacher, and his mechanical ingenuity enabled him to devise important improvements in the instruments used. At the age of twenty, while still attending Woodhouse's lectures, he set out to improve the chemist's blowpipe by providing a steadier stream of air than could be supplied by bellows or human lungs. The result of his efforts was his hydrostatic blowpipe, which relied on water pressure to insure a steady flow of the air introduced into a compartmentalized barrel by a bellows submerged at the bottom of the barrel (see fig. 2). Indeed, the machinery was so designed that two different gases could be admitted successively into the barrel and stored separately under pressure in airtight compartments. Thus, the operator could replace the stream of ordinary air from one compartment with a stream of pure oxygen from the other if he desired a greater intensity of heat for certain experiments, and he could do so much more conveniently than could be done with the gasometer Lavoisier had devised for producing an oxygen flame.

But Hare was not satisfied with producing a blowpipe that could rival the best European devices. Up to this time the fusion of refractory substances had been accomplished most efficiently by heating them over charcoal with an oxygen flame. But it was difficult to place the substance in the focus of the flame without interrupting the stream of oxygen that fed the flame. Moreover, the charcoal was consumed rapidly, and the substance undergoing fusion often ran into the pores of the charcoal and became difficult to observe. Reflecting on these difficulties, Hare tried to discover a way to make the upper surface of a substance heated over carbon as hot as the lower surface or to expose

[14] Brian Mason and H. B. Wiik, "The Composition of the Forest City, Tennasilm, Weston, and Geidam Meteorites," *Amer. Museum Novitates* 2220 (June 22, 1965). We are indebted to Professor John T. Wasson, professor of chemistry and member of the Institute of Geophysics and Planetary Physics, University of California, Los Angeles, for informing us of the existence of this modern analysis of the Weston meteor.

[15] *Medical Repository* 3 (1800) : p. 68.
[16] Robert Hare, *Memoir on the Supply and Application of the Blowpipe* (Philadelphia, 1802). Reprinted in *Philos. Mag.* 14 (1802–1803) : pp. 238–245, 298–306. See also *Annales de Chimie, ou Recueil de Mémoires Concernant la Chimie et les Arts qui en Dépendent, et Spécialement la Pharmacie* 45 (1802) : pp. 113–138.

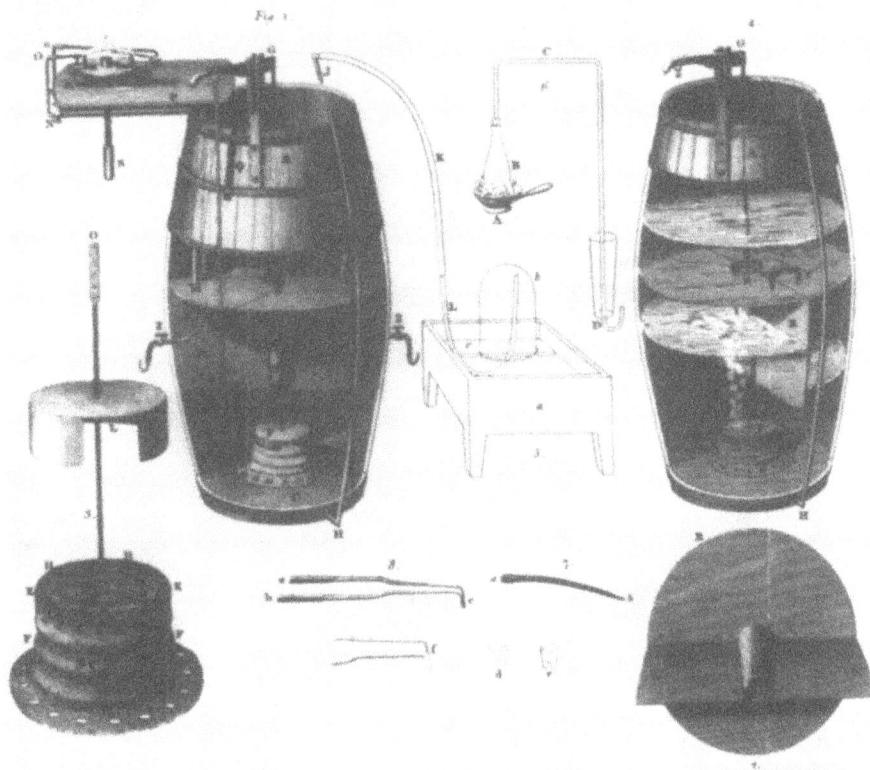

FIG. 2. Hare's hydrostatic blowpipe. Gas is drawn from the glass jar *b* inverted in the pneumatic trough *a* through the pipe *LKJIH* by operating the bellows *CF*. When the bellows is compressed the gas is forced into one of the compartments into which the cask is divided by the plate *EE* and stored there under hydrostatic pressure. Then another gas is drawn into the other compartment by turning the hood *F* of the bellows so as to direct the gas into that compartment for storage under pressure. Separate pipes *MNO* and *mno* conduct the gases from the two compartments to the plafform *P*, where stopcocks (*N* and *n*) are used to release the gases as desired at the point of combustion. Hare's complete description of the blowpipe and its operation may be found in Tilloch's *Philosophical Magazine* 14 (1802–1803) : pp. 238–245, 298–306, and plate VI at the back of the volume.

the substance on solid supports to a temperature even hotter than that generated by the union of porous charcoal with oxygen.

It soon occurred [Hare wrote] that these desiderata might be attained by means of flame supported by the hydrogen and oxygen gases; for it was conceived that, according to the admirable theory of the French chemists, more caloric ought to be extricated by this than by any other combustion.[17]

---

[17] Robert Hare, "Memoir on the Supply and Application of the Blowpipe," *Philos. Mag.* 14 (1802–1803) : p. 301. The word caloric (*calorique*) was coined by Antoine Lavoisier to designate the matter of heat. Caloric, he thought, penetrated the pores of all bodies, surrounding the particles and providing a repulsive force between them. Heating a solid increased the amount of caloric it contained, causing it initially to become a liquid and finally, with additional heating, to become a gas.

Hare's tightly compartmentalized blowpipe provided a means of storing the gases separately under pressure. The only problem was to keep them separate until the moment of ignition. For this purpose Hare conducted the gases in separate brass pipes which were fitted into tubular holes in a conical frustum of pure silver which converged at the point of ignition.

The pipes a b were then fitted into the mouths O, o, of the pipes of delivery [see fig. 2], so that the blow-pipe inserted into the larger hole of the frustum should communicate with the compartment containing the hydrogen gas, and that the other should communicate with that

---

Gaseous elements, e.g., oxygen and hydrogen, were supposed to contain a substantial amount of caloric; this explained why they released heat (caloric) during combustion or chemical combination.

which contained the oxygen gas. The cock of the pipe communicating with the hydrogen gas was then turned until as much was emitted from the orifice of the cylinder as when lighted formed a flame smaller in size than that of a candle. Under this flame was placed the body to be acted on, supported either by charcoal or by some solid and incombustible substance. The cock retaining the oxygen gas was then turned, until the light and heat appeared to have attained the greatest intensity. When this took place, the eyes could scarcely sustain the one, nor could the most refractory substances resist the other.[18]

Hare reported that this new device had enabled him to accomplish a complete fusion of barytes, alumina, and silex, producing in the case of barytes and alumina substances resembling white enamel and in the case of silex an ash-colored substance which sometimes exhibited brilliant yellow specks after long exposure. Lime and magnesia proved more difficult to fuse, partly because they were blown to one side by the flame, but in some instances small portions of these substances were converted into black vitreous masses when exposed on carbon to the oxyhydrogen flame. In other experiments platinum, gold, and silver were fused. "Had I sufficient confidence in my own judgment," Hare wrote, "I should declare, that gold, silver, and platina, were thrown into a state of ebullition by exposure on carbon to the gaseous flame; for the pieces of charcoal on which they were exposed became washed or gilt with detached particles of metal in the parts adjoining the spots where the exposure took place." After describing still other experiments on anthracite coal and various iron compounds, Hare rightly concluded that his invention had opened the way to experiments with a wide variety of substances previously considered infusible. "By means of the gaseous flame, such substances may be employed with the greatest facility in small analytical operations."

Hare's invention excited great interest among the chemists of Philadelphia. Elected to the American Philosophical Society at the age of twenty-two, Hare was induced to "reinstate" his apparatus and perform his experiments on the fusion of "earths" and metals again for the benefit of Woodhouse, Seybert, and the illustrious Dr. Priestley. Young Benjamin Silliman, who had recently arrived in Philadelphia to prepare himself to teach chemistry and natural history at Yale, assisted in the experiments. Silliman formed a warm friendship with the "genial, kind-hearted" Hare at Mrs. Smith's boarding house, where the two young men set up a chemical laboratory of their own in a spare cellar kitchen and bent their joint efforts toward improving Hare's recent invention.

Together they added strontianite to the list of hitherto infusible substances melted by the blowpipe, in addition to devising a method of combining the blowpipe with the pneumatic trough that greatly facilitated operations with large quantities of gases.

Silliman subsequently constructed an apparatus on this design at Yale College and reduced still further the list of substances incapable of fusion by the blowpipe. So impressed was Silliman with the importance of Hare's invention that he persuaded Yale to confer an honorary M.D. on Hare in 1806. Meanwhile Hare had reported the results of their joint experiments to the American Philosophical Society.[19]

Hare's election to the American Philosophical Society drew him into even closer contact with Woodhouse, Seybert, and Coxe, all of whom were very active in the affairs of the Society. Coxe seems to have devoted himself to medicine rather than chemistry or mineralogy on his return from Europe in 1796, but his *Philadelphia Medical Museum,* begun in 1805, provided a useful avenue of publication for chemical and mineralogical essays by Woodhouse, Seybert, James Cutbush, and others. Seybert, on the other hand, combined the study of medicine with an increasing devotion to chemistry and mineralogy during his extensive studies in Edinburgh, London, Paris, and Göttingen in the years 1793 to 1796. A paper entitled "Experiments and Observations on Land and Sea Air," based on samples he had collected during his return voyage, won him election to the American Philosophical Society in 1797. The following year he married Maria Sarah Pepper, daughter of a wealthy Philadelphia merchant. After trying medical practice without success for a year or two, he opened a pharmacy and chemical manufactory where he perfected a method of refining camphor and soon made a fortune from it. Meanwhile he remained active in the Chemical Society of Philadelphia and

----

[18] *Ibid.,* p. 303.

[19] Robert Hare, "An Account of the Fusion of Strontites, and Volatilitization of Platinum; and also of a New Arrangement of Apparatus," *Trans. Amer. Philos. Soc.,* O.S., 6, 1 (1809): pp. 99–105. This memoir was read before the Society on June 17, 1803; Benjamin Silliman, "On the Powers of the Compound Blowpipe," *Amer. Mineral. Jour.* 1 (1814): pp. 199–210. See also Fisher, *Life of Benjamin Silliman* 1: pp. 103–104. Silliman recalled that Hare's apparatus was "ingenious, but unsafe as regards the storage of the gases." "Novice as I was, I ventured to suggest to my more experienced friend that by some accident or blunder the gases—near neighbors as they were in their contiguous apartments—might become mingled, when, on lighting them at the orifice, an explosion would follow. . . . After my return to New Haven, I contrived a mode of separating these gases so effectually that they could not become mixed. Eventually I employed separate gasometers, one to contain the oxygen and the other the hydrogen, and during forty years that they were in use no accident ever happened." Silliman vigorously defended Hare's claim to priority in the invention of the compound blowpipe against the claims of Edward Clarke, Thomas Skidmore, and others. In 1839 Hare was awarded the Rumford Medal of the American Academy of Arts and Sciences for his technical contributions to the study of heat, including the oxyhydrogen blowpipe, the calorimotor, and the deflagrator. See also Donald McDonald, *The History of Platinum* (London, 1960), pp. 184–186.

the American Philosophical Society and devoted himself to the analysis of mineral waters.[20]

Seybert's interest and training in mineralogy and the cabinet of specimens he had brought back from Göttingen quickly made him the leading local authority on minerals. When Benjamin Silliman arrived in Philadelphia in the autumn of 1802 with Yale College's entire collection of minerals in a small candlebox, he was directed to Seybert as the expert who could identify the specimens. Meanwhile Seybert was busy adding American minerals to his collection. The extent of his collecting activity is revealed in a catalog of his collection in his own hand now in the collections of the Library of the Academy of Natural Sciences of Philadelphia. There are two such catalogs at the Academy, one of which bears the date 1825, the year of Seybert's death, on the cover. The other catalog appears from internal evidence to have been written considerably earlier. Some of the last entries (Nos. 1904–1951) concern specimens "collected on a Journey performed in the summer of 1810 from Philadelphia to the Niagara Falls & from Albany to Quebec." Presumably this catalog was turned over to the Academy of Natural Sciences when Seybert sold his collection to the Academy in 1812. A prefatory note at the beginning of the catalog explains:

This Catalogue is only intended to record localities and the names of minerals generally. A minute description of each specimen was not attempted, as this was reserved for a Catalogue formed upon the principles of Haüy's System of Mineralogy. Experience has taught me the utility of the present record of localities, for the places whence the specimens were brought are soon forgotten and thus the utility of them in great measure is lost, especially when they are intended as facts whereon to build a system of Geology of a particular country or district.[21]

---

[20] Grady Roney and J. H. Hoch, "Adam Seybert (1773–1825)," *Amer. Jour. Pharmacy.* **127** (1955): pp. 346–358; Caldwell, *Autobiography,* pp. 309–311; E. F. Smith, *Chemistry in Old Philadelphia* (Philadelphia, 1919), pp. 24–43.

[21] Catalogue of Minerals, Library of the Academy of Natural Sciences of Philadelphia, prefatory note. This catalog, written in Adam Seybert's hand, was erroneously attributed to Gerard Troost and included in Collection 372 at the Library of the Academy. It is undated, but internal evidence suggests strongly that it was completed shortly before Seybert sold his collection in 1812. The numbers of the entries do *not* correspond to those of the minerals in the Seybert collection at the Academy; these are tagged in correspondence with the entries in a later catalog in Seybert's hand dated 1825. It is not clear why only some, not all, of the minerals entered in the earlier catalog can be found in the later one. Perhaps Seybert withheld some of the minerals listed in the earlier catalog, or perhaps some of the specimens were discarded or removed from the collection after it was purchased. Considerable changes must have taken place in the collection between 1812 and 1825, since the total number of specimens listed in the final catalog was 157 fewer than the number listed in the earlier one despite the addition of many new minerals after 1812.

This early catalog is especially interesting as a detailed record of where and how Seybert obtained his American specimens, which constituted a sizable part of the collection. Among the first one hundred specimens (mostly quartzes), six-tenths were from America, including occasional specimens from Quebec, Mexico, and South America. Among those from the vicinity of Philadelphia are lamellar quartz from "the lane which leads from the great road on Chesnut [sic] Hill, to Paul's mill, Wissahickon"; hackley quartz "from Pawling's & Dill's mine near Perkiomen Creek, Montgomery Co. Penn."; the pyramidal termination of a prismatic quartz crystal plowed up in the fields in Berks County; and granular quartz "found on the surface on Coates's farm, 14 miles west of Philada. near the Lancaster Road."

From farther afield came prismatic quartz with pyramidal terminations "from the South mountain near Egges Iron forge"; "very transparent prismatic quartz, considerably flattened," found "on the surface near Nazareth, Northampton Co. Penn"; "crystallized quartz consisting of an hexahedral prism terminating at each end with six distinct pyramids" from Virginia; and various specimens from Mexico, Brazil, and "South America." Items 538 and 539 were fragments of the Weston meteor of 1806, contributed by Benjamin Silliman. Items 449 and 450 were basalts from the supposed "wall" in North Carolina concerning which Woodhouse had written. The European quartzes included specimens from Iceland, Bohemia, Saxony, England, Scotland, the Alps, Hanover, Tuscany, and Spain.

The specimens contributed by Captain Lewis of the Lewis and Clark expedition are the most interesting of all. One of these, a sulphuret of iron containing gold, was from Virginia, but the remaining thirty-four were collected on the famous expedition to the Pacific. How Seybert acquired these specimens is not indicated. Most of them were collected along the banks of the Missouri River—quartz crystals, pipeclay, agatized wood, sulphate of lime, fossil shells, etc. There was also pumice from the Pacific Ocean, green clay "from the Kooskoosche River, west of the Rocky mountains," magnetic iron sand from the borders of the Pacific Ocean "near to the mouth of Columbia River," and the like.[22]

---

[22] The full list of specimens contributed by "Captn. Lewis" is as follows:

No. 119: "Fortification agate with small quartz crystals, from the Missouri, brot. by Captn. Lewis."

Nos. 143 and 144: "Polienschiefer. On boarders [sic] of the Missouri, August 22d. 1804. by Captn. Lewis."

No. 149: "Pumice. Pacific ocean. Captn. Lewis."

No. 179: "Agatized wood. in the bluffs near fort Mandan. Captn. Lewis."

No. 366: "Pipe Clay, south bank of the Missouri River. Captn. Lewis."

No. 413: "Green Clay. from the Kooskoosche River, west of the Rocky mountains. Captn. Lewis."

All in all, Seybert's collection was an interesting and valuable one, to which he continued to add throughout his life. Discussing the importance of Seybert's contributions, Clifford Frondel, professor of mineralogy in the Department of Geological Sciences at Harvard University, writes:

I have looked over Seybert's catalogue. It is not the ordinary kind of mineral collection catalogue, comprising a sequential list of accessions in essentially chance order. As the preface implies, and the contents indicate, it is instead an inventory of an organized collection already in being. At least this is true up to about no. 1700. Following entries seem to be addenda, entered out of order, and represent specimens, groups of specimens and small

No. 534: "Pumice, found floating on the Missouri. Captn. Lewis."
No. 535: "Slag like Lava. in the sides of the hills in the neighborhood of fort Mandane [sic] 1609 miles miles above the mouth of the Missouri. Captn. Lewis."
No. 536: "Lava found near fort Mandane [sic] 1609 miles above mouth of the Missouri. Captn. Lewis."
No. 748: "Fossil shells from banks of Missouri, the stratum superimposed on a stratum of Sandstone. Captn. Lewis."
No. 750: "Fossil Shell—found at the upper point of the big bend of the Missouri. Captn. Lewis—perhaps from the compartments of a large cornum ammonis."
Nos. 796–798: "Regular crystals of sulphat of Lime. banks of the Missouri. 23d Augst. 1804. Captn. Lewis."
No. 799: "Crossed crystals of Sulphat of Lime. locality same as 798 &c."
Nos. 800–801: "Crystallized sulphat of Lime. Calumet Bluffs. Missouri. Captn. Lewis."
Nos. 802–805: "Crystallized sulphat of Lime. Locality same as 801."
No. 814: "Sulphat of Lime with carbonat of Lime, found above white chalk bluffs on the Missouri. Captn. Lewis."
No. 904: "Keffekill. found at the Wallenwaller [sic] nation on Columbia River. Captn. Lewis." [Keffekill is an old name for an impure clay.]
No. 1115: "Muriat of Soda. found adhering to rocks, thro' which a salt fountain issues on the south side of the southern branch of the Arkansas River. This salt gave rise to erroneous report of a Salt Mountain being found in Louisiana. Captn. Lewis."
Nos. 1127–1129: "Fibrous sulphat of alumine. near alum bluffs on the Missouri. 21st. Augst. 1804. Captn. Lewis."
No. 1134: "Sulphuret of Iron in a state of decomposition with an efflorescence of sulphat of Iron on the surface. banks of the Missouri. Captn. Lewis."
No. 1135: "Sulphuret of Iron decomposed with efflorescence of Sulphat of Iron—found with the carbonated wood 1169. Missouri. Captn. Lewis."
No. 1138: "Aluminous earth. Missouri near the mouth of rapid river. Captn. Lewis."
No. 1169: "Carbonated wood. Missouri. Captn. Lewis."
No. 1377: "Globular sulphuret of Iron. at the entrance of Cannonball river on the Missouri. Some specimens of many lbs. weight. Captn. Lewis."
No. 1409: "Magnetic Iron sand, borders of the Pacific ocean near the mouth of Columbia river. Captn. Lewis."
(No. 1203 is from Captn. Lewis but not collected on the expedition: "Sulphuret of Iron containing gold. Virginia. Captn. Lewis.")

collections that were acquired in various ways but certainly in part from trips.

The collection evidently was not scientific in purpose but was primarily intended as a utilitarian guide to the common minerals, ores, ceramic materials and rocks. The labelling of the specimens, the manner of reference to authority and especially the crystallographic statements indicate that Seybert was a practical rather than a professional mineralogist, using European standards of the time.[23]

Seybert's zeal for mineralogy and his ambition to achieve distinction in this field are shown in his letters to Robert Gilmor, Jr. of Baltimore and to Colonel George Gibbs of Newport, Rhode Island, both of whom, like Seybert, had acquired a taste for mineralogy during their travels abroad. In April, 1807, Seybert wrote to Gilmor requesting the loan of Brochant's Traité élémentaire de minéralogie, explaining:

It is my intention to translate the work into English provided I think it will suit our students. My collection of minerals is arranged and makes a tolerable show, it is my wish to do something in this way the ensuing summer and it is probable I may accumulate sufficient matter to add in the form of notes by way of commencing an American Mineralogy. . . .
Any information which you will please to communicate on the subject of American Mineralogy, especially if specimens can accompany the same, will be received with great satisfaction.[24]

In February of the following year Seybert sent some minerals to Colonel Gibbs, who was departing for Europe, and requested him to find him a mineralogical correspondent in Paris.

Pray ask Haüy, Lamethrie [sic] & any other you please if a correspondence in this way would be agreeable. If whilst you are abroad it should be perfectly convenient for you to procure some specimens for me I will esteem it as a very particular favour. Will you be so good & purchase for me at Paris a set of Haüys models of the integrant molecules & set of primitive figures of crystals—also a goniometer—a small platina crucible & an agate mortar fit for analyzing minerals. Also Chaptals Treatise I believe entitled "the Application of Chemistry to the Arts" or something like this. 2 graduated detonating Tubes, with brass conductors & chain for firing gases. The Instrument described by Hauy Plate 8 fig 76 for ascertaining the electricity of minerals by friction & heat. Also the instrument of Nicholson, described in the same plate fig. 75, for ascertaining the specific gravity of minerals. Your attention to these objects will be conferring a great favour on me. When the minerals are forwarded to you I will advise you of the stage by which they are sent.[25]

[23] Letter from Clifford Frondel to John C. Greene, Cambridge, Mass., June 13, 1973. The authors are much obliged to Professor Frondel for permission to quote from this letter.
[24] Adam Seybert to Robert Gilmor, Jr., Philadelphia, April 1, 1807, Dreer Collection—Physicians, Surgeons & Chemists—4, n. p. Historical Society of Pennsylvania.
[25] Adam Seybert to Colonel George Gibbs, Philadelphia, February 25, 1808, Gibbs Correspondence, Newport Historical Society.

Apparently Gibbs was very obliging, for in September, 1808, Seybert welcomed him back to America and thanked him for the items he had shipped to Philadelphia.

> . . . the articles you forwarded came very à propos, for you must know, I contemplated delivering a course of lectures this winter! Your assistance would be of the greatest consequence to me.[26]

But Seybert's needs were still not satisfied. He asked Gibbs whether he had duplicates of Haüy's primitive forms and molecules or models of de l'Isle's figures and, if not, whether it might be possible to make copies in plaster of Paris. If this was not feasible, would Gibbs give him the name of a reliable firm in Paris where he could procure these items? In the same month Seybert held out to Robert Gilmor the prospect of "a chain of correspondence from north to south that would provide materials for constructing a mineralogical chart of the United States" and announced that he had been offered the professorship of chemistry in the newly established College of Medicine of Maryland.[27]

The first fruits of Seybert's researches on American mineralogy were published in Coxe's *Philadelphia Medical Museum* in July, 1808, under the caption "A Catalogue of some American Minerals, which are found in different Parts of the United States. By ADAM SEYBERT, M.D., member of the American Philosophical, and of the Royal Societies of Göttingen." In his prefatory remarks Seybert explained that he had been induced to publish this catalog by "the inquiries which are daily made by many of my fellow-citizens, and from the numerous applications of foreigners, who desire to have specimens of American minerals." The science of mineralogy, he declared, had recently been placed on a firm footing in Europe, but had been "almost totally neglected" in the United States.

The Dutch naturalist Gronovius had described some American minerals in his *Index supellectilis lapideae* of 1740, but vaguely and without a clear indication of the localities from which they came. By publishing a more ample and precise list Seybert hoped to encourage others to join the effort. If those who collected minerals were careful to note their geological situations, the groundwork would be laid for the eventual publication of a correct geological map of the United States.

Seybert then proceeded to describe forty American minerals from the specimens in his cabinet, giving the names assigned to each of them by three eminent European mineralogists—Richard Kirwan, René Just Haüy, and Abraham Gottlob Werner—and indicating briefly their characters and localities, with occasional additional information concerning the occurrence of the same minerals elsewhere, without, however, indicating the sources of this information. His account of serpentine shows that he was interested in geology as well as mineralogy:

> This mineral occurs abundantly in an amorphous independent state, in Chester, Montgomery, and Delaware counties, Pennsylvania. It is frequently veined, and in many instances is variously coloured.
>
> It has been discovered near Newport, Rhode Island, and the perpetual green of the hills of the state of Vermont is owing to their being formed of this rock, which, in Pennsylvania, is well known by the name of *barren stone*.
>
> Serpentine rock is common at Hoboken, New Jersey.
>
> All the above serpentines attract the magnetic needle, though in a feeble manner, except that from the neighbourhood of Newport, which is the only specimen that exhibited, in every evident manner, two *distinct poles*.
>
> It seems probable to me that this rock commences in Rhode Island; that it assumes a south-western course, and traverses Connecticut, New York, New Jersey, Pennsylvania; and thence passes to an unascertained distance; that it is one continued chain: it closely follows the direction of our principal mountains.[28]

Seybert's descriptions are mostly brief, but there are occasional interesting observations, as when he mentions seeing a specimen of celestite collected near Frank's Town, Pennsylvania, by a German mineralogist named Schuetz in the collection of Professor Johann Friedrich Blumenbach of the University of Göttingen.

Meanwhile Seybert had been asked to assist in arranging the mineralogical specimens that had been accumulating at the American Philosophical Society for many years. The Society's natural history collection, including minerals, was housed in Philosophical Hall in a room leased to the College of Physicians, and on May 5, 1797, Woodhouse and Seybert were asked to make a list of the minerals in the collection. The number of specimens was not large, but a substantial addition was made in 1802 when the collection of Thomas P. Smith was bequeathed to the Society.

Self-educated in science, Smith had been active in both the Chemical Society of Philadelphia and the American Philosophical Society and had published several articles in the Society's *Transactions,* including an account of the basalts of the Conewago Hills. In the years 1800–1802 he made a "mineralogical tour" of Europe, visiting Britain, France, Switzerland, and Scandinavia, keeping extensive notebooks concerning mining and manufacturing establishments in

---

[26] Adam Seybert to Colonel George Gibbs, Philadelphia, September 17, 1808, Gibbs Correspondence, Newport Historical Society.

[27] Adam Seybert to Robert Gilmor, Jr., Philadelphia, September 5, 1808, Etting Collection—Physicians, p. 82, Historical Society of Pennsylvania.

[28] Adam Seybert, "A Catalogue of Some American Minerals, Which Are Found in Different Parts of the United States," *Philadelphia Medical Museum* 5 (1808) : p. 266. The article is in two installments, pp. 152–159, 256–268.

these countries, and collecting rocks and minerals "in a geological & manufacturing point of view" for the purpose of promoting American economic development. On the return voyage, however, he was mortally wounded in the accidental explosion of one of the ship's guns. He died at sea, but not before he had had time to make a will bequeathing his collections and notebooks to the American Philosophical Society. Soon afterward more than fifteen boxes and trunks containing these materials were delivered to the Society, and Woodhouse, Seybert, Coxe, Hare, and Benjamin Smith Barton were appointed a committee to examine the specimens and put them in scientific order.[29]

On behalf of the committee Robert Hare wrote to Robert Gilmor, Jr., in Baltimore for advice concerning the kind of case suitable for displaying Smith's specimens. The son of a wealthy Baltimore merchant, Gilmor had developed an interest in mineralogy during his grand tour of Europe in 1800–1801. In Paris he had been deeply impressed by the immense collections of minerals and plants at the Jardin des Plantes. "I could spend a year in Paris ... extremely well," he wrote his brother, "in attending the various lectures which are publicly given by the first characters in almost every branch of Science, and cultivating the society of the natives, which however is not easily obtained." He did not say whether he had attended the Abbé Haüy's lectures on mineralogy but he continued to visit natural history cabinets as he traveled through Italy, Germany, Holland, and England and to purchase objects of scientific and cultural interest, including mineral specimens. "I believe I have not misemployed my time, nor trifled it away in unnecessary shew or mere amusement," he wrote his brother from London. "The materials I have collected both for my improvement in art & science will I trust be a rich fund for my friends as well as myself to draw pleasure & instruction from."[30] Apparently these materials included a handsome cabinet of minerals, for Hare addressed him as a person knowledgeable about mineralogical collections.

Few are as well aware as you of the importance of such an establishment. To you it is well known how little the

department of mineralogical science had been explored or understood by our countrymen. In fact America boasts not a single school for this Science![31]

Despite whatever advice Gilmor may have offered, the mineralogical committee made slow progress in arranging the Society's collection. One suspects that personal antipathies as well as scientific disagreements made it difficult for the members to agree on the proper method of arrangement. Even a directive from the Society instructing them to adopt the system set forth in the second edition of Richard Kirwan's *Elements of Mineralogy* failed to produce harmony. Early in 1805 the committee was disbanded, and Seybert was given sole responsibility for the task. Meanwhile new specimens were being added to the collection. President Jefferson sent a box of earths and minerals collected by Captain Meriwether Lewis, and Robert R. Livingston donated volcanic specimens from Naples. Juan Manuel de Ferrer sent specimens from Vera Cruz. Other Mexican minerals came from Dr. Samuel Brown in New Orleans, and William Maclure donated a case of minerals and rocks collected in Spain.[32]

For reasons that are not entirely clear, Seybert asked to be relieved of the duty of arranging the collections in November, 1808, and retired from the mineralogical committee. Fortunately for the Society a young French mineralogist, Silvain Godon, appeared in Philadelphia about this time to put the mineral specimens in order. Godon's arrival in America had been noted in July, 1807, by François André Michaux in a letter to Benjamin Vaughan, brother of John Vaughan, the treasurer of the American Philosophical Society. Writing from Boston, Michaux said:

A young Frenchman M. Godon St. Memin, whom I knew at Paris has just arrived here on the last ship, which brings the latest news from Europe. Mineralogy is for him what botany is for me. He is certainly one of the ablest mineralogists who have ever come to this country where he counts on locating, being very discontented with France. He is laden with letters of introduction, principally for the middle and southern states where he plans to go. He has one, I think, for Mr. John Vaughan.[33]

Soon Godon arrived in Philadelphia bearing letters of introduction from William Maclure and immediately undertook a mineralogical excursion to Maryland, publishing the results in the *Transactions* of the American Philosophical Society under the title "Observations to Serve for the Mineralogical Map

[29] See Smith, *Chemistry in Old Philadelphia*, pp. 14 ff. for a brief biography of Thomas Peters Smith. See also *Trans. Amer. Philos. Soc*, O.S., 4 (1799) : pp. 431, 445; "Early Proceedings," pp. 116, 120, 203–204; *Medical Repository* 3 (1800) : pp. 151–154, 253–257. Smith's notebooks are at the Library of the American Philosophical Society.

[30] Robert Gilmor, Jr. to William Gilmor, Paris, June, 1800, letter books of Robert Gilmor, Gilmor Papers, letter No. 13, Maryland Historical Society; same to same, London, August 25, 1801. The Gilmor Collection (387) also has a printed "Memoir, or Sketch of the History of Robert Gilmor of Baltimore" with a prefatory handwritten note by Robert Gilmor, Jr., dated Baltimore, December 25, 1840. See also Francis C. Haber, "Robert Gilmor, Jr.—Pioneer American Autograph Collector," *Manuscripts 7* (Fall, 1954) : pp. 13–17.

[31] Robert Hare to Robert Gilmor, Jr., Philadelphia, August 28, 1803, Etting Collection—Scientists, Historical Society of Pennsylvania, p. 37.

[32] "Early Proceedings of the American Philosophical Society," pp. 379, 382, 399, 401, 416.

[33] François André Michaux to Benjamin Vaughan, Boston, August 1, 1807, Benjamin Vaughan Papers, American Philosophical Society. Original in French.

of the State of Maryland." "Specimens of most of the minerals mentioned in this Memoir," Godon noted, "are deposited in the Collection of the Philosophical Society of Philadelphia."[34]

After returning to Boston for a year to deliver a course of public lectures and to geologize in eastern Massachusetts and the District of Maine, Godon settled in Frankford, near Philadelphia, and was soon busily occupied in various scientific and business ventures. On the business side he undertook the manufacture of yellow pigments derived from a combination of lead with chromic acid, the latter obtained from deposits of chromate of iron near Baltimore.[35] Beginning in 1808, Godon gave several courses of public lectures on mineralogy, eventually gaining permission from the American Philosophical Society to use its collections to illustrate his lectures in return for his services in arranging the collections. Elected to the Society and to the Linnaean Society of Philadelphia, he began to lay plans for publishing by subscription *A Treatise on Mineralogy Adapted to the Present State of Science, Including Important Applications to the Arts and Manufactures.*

At this juncture, when the prospects for mineralogical science in Philadelphia seemed to be brightening, a series of unhappy events cast a shadow over these prospects. In the summer of 1809, Woodhouse's promising career was brought to a premature end by a fatal stroke. His collection of minerals, including those he had acquired on a trip to London in 1802, went to the American Philosophical Society, but the

chemical society he had organized and led expired with its founder, and bitter rivalry broke out for the highly coveted chemical chair at the School of Medicine, held by Woodhouse since 1795. Once again Seybert, Coxe, and Hare were candidates for the position. Once again Benjamin Rush's influence secured the appointment of his favorite, this time John Redman Coxe. Disappointed in his fondest hopes, Hare continued to pursue chemistry as an avocation while making a living as a brewer, meanwhile maintaining a brisk correspondence with his friend and former collaborator Benjamin Silliman. Seybert was furious at the "Jesuitical conduct" of Benjamin Rush, as he described it to Colonel Gibbs.

Seybert's devotion to mineralogical pursuits had suffered a setback even earlier, for he wrote Gibbs in February, 1809:

My attention will hereafter be taken up in a great measure as a manufacturer of cotton. I shall not totally neglect mineralogy & I intend to have a laboratory fitted up the next summer to enable me to pursue the science of chemistry.[36]

The appointment of Coxe to the chemical chair at the University drove Seybert even farther afield from science. When the congressional seat from the Philadelphia district fell vacant in 1811, Seybert ran for the office on the Democratic ticket and won the election. From 1811 to 1815 and again from 1817 to 1819 he served in Congress, finding little time for the study of science, as he confessed to Parker Cleaveland in 1813 when Cleaveland asked him for mineralogical information.

It would give me great pleasure to be able to answer the several queries which you have communicated in your letter. For the last four years my attention has been taken from subjects of mineralogy; the few facts which I did possess, have in a great measure escaped me. The knowledge which you desire from me, should depend upon facts alone. I have no collection or notes to refer to. It would be painful to me to lead you into any errors on this subject.[37]

In the summer of 1812 Seybert had sold his mineral cabinet containing nearly 2,000 specimens to the newly formed Academy of Natural Sciences of Philadelphia.

Godon, too, suffered reverses that destroyed his ambition to achieve a reputation in mineralogy. The details of his misfortunes are not known, but in December, 1812, he was in prison, presumably on account of debts incurred in his business adventures. From the Philadelphia Prison he informed the American

---

[34] Silvain Godon, "Observations to Serve for the Mineralogical Map of the State of Maryland," *Trans. Amer. Philos. Soc.,* O.S., 6, 2 (1809) : pp. 319–323.

[35] The information that Godon was engaged in the manufacture of chromic yellow is from Bruce's *Amer. Mineral. Jour.* 1 (1814) : p. 125. Godon's interest in deriving pigments from chrome dated back to his Parisian days. See "Sur La Belle Couleur Verte que le Chrome Peut Fournir à la Peinture," *Annales du Muséum National d'Histoire Naturelle* 4 (1804) : pp. 238–241. Thomas Jefferson was acquainted with Godon's activities. In a letter to Charles Willson Peale dated Monticello, May 5, 1809, he remarked : "I have communicated to my grandson our consent to his attending Mr. Godon's lectures in mineralogy till the botanical course ends, after which he is to return home." (*Penna. Mag. History and Biography* 28: pp. 318–319). On April 19, 1816, Jefferson wrote to Governor Wilson C. Nicholas, who had requested the name of someone who could undertake a "mineralogical survey" of Virginia : "I have never known in the United States but one eminent mineralogist, who could have been engaged on hire. This was a Mr. Goudon [*sic*] from France, who came over to Philadelphia six or seven years ago. Being zealously devoted to the science, he proposed to explore the new field which this country offered; but being scanty in means, as I understand, he meant to give lectures in the winter which might enable him to pass the summer in mineralogical rambles. It is long since I have heard his name mentioned, and therefore do not know whether he is still at Philadelphia, or even among the living." See A. A. Libscomb and A. E. Bergh, eds., *The Writings of Thomas Jefferson* (20 v., Washington, D.C., 1907) 14 : p. 486.

[36] Adam Seybert to Colonel George Gibbs, Philadelphia, February 17, 1809, Gibbs Correspondence, Newport Historical Society; Seybert's comments on Rush's conduct are in a letter to Gibbs, Philadelphia, July 21, 1809.

[37] Adam Seybert to Parker Cleaveland, Washington, December 16, 1813, Cleaveland Correspondence, Division of Manuscripts, New York Public Library.

Philosophical Society of the sale of his effects and implored his fellow members not to forget that it was he "who rescued your collection from the frightful chaos in which the stones were." Even in prison, he wrote, he was engaged in designing a plan for reorganizing the Society to make it more effective in the cause of science. In the meantime he offered the Society the few pieces of chemical apparatus still in his possession as a token of his esteem.[38] Despite these brave words, Godon was never again to play a significant role in Philadelphia science. The records of that city's scientific societies show little trace of him after 1812 except for brief references to the disposition of his collection of minerals and the terse notice of his death in the minutes of the American Philosophical Society for November 6, 1840: "Mr. Du Ponceau announced the death of Mr. Sylvanus Godon, a member of this Society." Apparently Godon had continued to collect minerals, for a rough note with the marginal notation "1840?" in the Patterson papers at the American Philosophical Society places a high value on his cabinet, which had been offered for sale. Godon's continuing esteem for the Society is shown by the gift of a small collection of artifacts from the Hawaiian and the Fiji Islands made by his son, Sylvanus William Godon, who was to have a distinguished career in the United States Navy.[39]

Mineralogy made little progress in the School of Medicine during the tenure of John Redman Coxe in the chemical chair. Coxe was more interested in medical and humanistic studies than in the physical sciences. Although he urged his chemistry students to collect American minerals, he failed to inspire them by his own example, preferring instead to buy specimens in England at considerable expense. In 1819 he boasted to Parker Cleaveland that few mineral collections in America could rival his in size and beauty of specimens, but his cabinet seems to have played no significant role in promoting the study of mineralogy in Philadelphia. When the University of Pennsylvania appointed Thomas Cooper professor of mineralogy and chemistry in a newly established Faculty of Physical Science and Rural Economy in 1816, Coxe was embittered and chagrined. In 1818 he exchanged his professorship of chemistry for the professorship of materia medica, precipitating another struggle for the chemical chair, to which Robert Hare was finally appointed after months of acrimonious debate. By this time Hare had acquired considerable interest and competency in mineralogy through his contacts with Benjamin Silliman, but he did not devote much attention to this science during his long and distinguished tenure as professor of chemistry in the School of Medicine. There, as elsewhere, the connection between mineralogy and medicine proved too remote to provide a secure foundation for mineralogical science.

Thomas Cooper's appointment as professor of mineralogy and chemistry in the undergraduate college might well have furnished a much-needed stimulus to the study of mineralogy and geology if the University had been able to support the appointment with adequate funds, as Yale College had done for Benjamin Silliman. A redoubtable little man who impressed even Thomas Jefferson by the breadth and depth of his learning, Cooper had taught chemistry, mineralogy, and geology at Dickinson College in Carlisle, Pennsylvania, for three years, had contributed to Bruce's *American Mineralogical Journal*, had edited an American edition of Frederick Accum's *System of Theoretical and Practical Chemistry*, and had devoted considerable attention to metallurgy and practical chemistry in his role as editor of *The Emporium of Arts and Sciences*, which he took over from Coxe in 1813. The first volume of this journal under Cooper's editorship (1813) was chiefly devoted to the manufacture of iron and steel, the third volume to copper, brass, and tin manufacture, with sections on the chemical analysis of these ores. Cooper's remarks were interspersed with information borrowed from a wide variety of European writers, practical and theoretical. "I wish I could have made the preceding papers more popular and less chemical, but I could not," Cooper wrote. "The time is at hand when young iron masters will find it necessary to become acquainted with chemistry. . . ."

The elements of natural philosophy and of chemistry, now form an indispensable branch of education among the manufacturers of England. They cannot get on without it. They cannot understand or keep pace with the daily improvements of manufacture without scientific knowledge: and scientific knowledge is not insulated; it must rest upon previous learning. The tradesmen of Great Britain, at this day, can furnish more profound thinkers on philosophical subjects, more acute and accurate experimenters, more real philosophers thrice told, than all

---

[38] Silvain Godon to the American Philosophical Society, Philadelphia Prison, December 11, 1812, Archives of the American Philosophical Society. Apparently Godon did succeed in arranging the American Philosophical Society's collection of minerals, for Dr. James Mease reported in his *Picture of Philadelphia* (Philadelphia, 1811), p. 301, that the Society had "an extensive collection of minerals from all quarters of the world, scientifically arranged." Further light on Godon is found in the autobiographical notes of Dr. John Collins Warren of Boston, published in Edward Warren's *The Life of John Collins Warren, M.D.* . . . (2 v., Boston, 1860) 2: p. 10: "Godon was the son of a wine-merchant in Paris. Marrying against his father's consent, he fled his country, and came here [Boston]. . . . Afterwards he went to Philadelphia; became insane; passed a number of years in the hospital, and finally died there." For Godon's Boston visit, see below p. 70.

[39] Godon died October 27, 1840. Concerning his son Sylvanus William Godon, see "Early Proceedings of the American Philosophical Society," p. 655; *Appleton's Cyclopedia of American Biography* (7 v. New York, 1888–1900) 2: pp. 670–671.

Europe could furnish a century ago. I wish that were the case here; but it is not so. I fear it is not true, that we are the most enlightened people upon the face of the earth; unless the facility of political declamation be the sole criterion of decision, and the universal test of talent. We should greatly improve, in my opinon, by a little more attention to mathematical and physical science; I would therefore encourage whatever would introduce a general taste for such pursuits.[40]

To illustrate the utility of chemistry in agricultural pursuits Cooper reported his analysis of nine samples of magnesian limestone collected in Chester County by Richard Peters, president of the Philadelphia Society for Promoting Agriculture, comparing his own methods of analysis with those used by European chemists. In the third volume of the *Emporium* he gave an "Outline of Geology," drawing on Werner's views as expounded by Robert Jameson and illustrating his account with references to mineral localities in Europe and America.[41]

Meanwhile Cooper kept in touch with the scientific community in Philadelphia and elsewhere and looked for opportunities to improve his professional prospects. On July 19, 1815, he informed Colonel Gibbs that he was planning to leave Dickinson College: "Neither the emoluments or the discipline suit me; & a more decided blockhead than our principal (Dr. Atwater) cannot be found, unless after very diligent search." In December he wrote Gibbs from Philadelphia:

. . . I quitted Carlisle, and finding no chemical situation that would afford me a reasonable livelihood I determined to pursue my profession as a lawyer at New Orleans. But repeated attacks of catarrh and influenza, with depletion that brought on a month's gout, confined me nearly two months. As soon as I could walk about the room, I went to Baltimore to embark for New Orleans, but just as I was about to put my luggage on board, the death of Dr. [Benjamin Smith] Barton of this City, made it

so probable that changes would take place in the chemical chair, that I was induced to comply with the invitations of my friends, and try my chance here. Whether I shall obtain it, is very uncertain: principally owing to my not having graduated as a Physician; though it is acknowledged I have practiced successfully for twenty years in this country; but irregularly, as I *never* took a fee. Had I foreseen this objection, I could have obtained an honorary degree long ago.

However, succeed or not in my present object, I shall stay here and give chemical lectures. While my determination was to go to New Orleans, I forebore trespassing on your kindness for the remainder of the minerals which you were so kind as to offer to me. Nor till I saw the means of using them usefully according to your intentions, should I have troubled you on the subject. But here, I propose to use them for the purpose of instruction if I can, and therefore take the liberty of renewing my request. . . .[42]

Soon afterward Cooper was appointed professor of mineralogy and chemistry at the University of Pennsylvania and began to plan his lectures on geology and mineralogy. A syllabus of these lectures published in 1817 reveals the essentially geological approach Cooper adopted in the study of mineralogy:

. . . the study of isolated minerals, independent of their natural situations, is the least important, and therefore the least interesting part of the knowledge which the sciences of mineralogy and geology, hitherto improperly divided, are calculated to teach. What with the beauty of many minerals, the delicacy of their colours, the varieties of the crystallization, and other external characters, cabinet collectors are led, or rather misled, to pass without notice the more common minerals, to neglect the great outlines of the science as nature offers them to our view, to confine themselves to minerals, so scarce, valuable and beautiful, and to make a cabinet rather an ornament and a plaything than an instrument of illustrations; the chrystaglognosts in particular look at a mineral with an eye so microscopic that they seem to merge all ideas of utility in collecting and tracing the form and shape of the mineral they subject to examination, which when known is of very dubious utility, unless as an adjunct to confirm, or explain in a few instances comparatively, the indications of other external characters, or the information presented by chemical analysis. . . .

The mode of teaching mineralogy that I have chosen to adopt is to make it consequent upon, and secondary to, geology. What I want to see is that when our young men travel over any part of this vast continent, they know at sight the ground they are upon and form a reasonable conjecture of what the earth underneath them contains from the nature of the surface they tread under their feet. . . .

It is thus, if at all, that the mineral riches of our country will be gradually searched out and made known. . . . I have for this purpose endeavored, not always indeed with perfect success, so to form my collection as to show, not only the gangue—the kind of stratum in which the

---

[40] Thomas Cooper, "Prospectus of the Emporium of Arts and Sciences," *The Emporium of Arts and Sciences*, n. s., 1 (1813): p. 6. Unfortunately, Thomas Cooper's scientific career has never been studied carefully. There is a brief appreciation of his work in Smith, *Chemistry in Old Philadelphia*, pp. 62 ff. For an idea of the range of Cooper's interests and erudition see *The Introductory Lecture of Thomas Cooper, Esq., Professor of Chemistry at Carlisle College, Pennsylvania . . . With Notes and References* (Carlisle, Pennsylvania, 1812), in which Cooper describes mineralogy as "the handmaid, the short hand of Chemistry."

[41] Thomas Cooper, "Outline of Geology, by the Editor," *The Emporium of Arts and Sciences*, n. s., 3 (1814): pp. 412–444. "I wonder whether our state legislatures will ever become sufficiently enlightened to appoint persons to explore each state mineralogically," Cooper wrote to Colonel Gibbs on April 2, 1815. "We are sadly at a loss in this country for Tin and Antimony" (Gibbs Correspondence, Newport Historical Society). According to Peter A. Browne, in his *Address Intended to Promote a Geological and Mineralogical Survey of Pennsylvania . . .* (Philadelphia, 1826), p. 3: ". . . so late as the year 1811, the mineralogical excursions of the then Judge, now Professor Cooper, along the Susquehanna, brought upon him the charge of lunacy."

[42] Thomas Cooper to Colonel George Gibbs, Philadelphia, December, 1815, Gibbs Correspondence, Newport Historical Society; see also Cooper's letters to Gibbs dated April 2 and July 19, 1815.

minerals are imbedded—but, as much as possible, the minerals found in the immediate vicinity.[43]

These lectures, Cooper wrote to Benjamin Silliman, would begin in November, 1817. A printed syllabus announced that Cooper planned to discuss "the means of distinguishing mineral substances artificially; by the file, the knife, the blow-pipe, the mineral acids; by their crystallization and the goniometer; their phosphorescences, their refraction, their magnetism and polarity, their electricity, specific gravity, etc." and to review the mineral classifications of Werner, Jameson, Thomson, Kirwan, Babington, Kidd, Clarke, Romé de l'Isle, and Haüy. "When the substances are exhibited *in* the rocks, as they are actually found in nature, with their accompanyments," the syllabus concluded, "they convey, as I think, far more precise information than any hand specimen or cabinet collection can possibly afford. For the motto of mineralogy is that of man in the civilized world; *noscitur à socio* 'shew me your companions and I will tell you who you are.'" According to an advertisement in the *Analectic Magazine*, the lectures were to be illustrated by appropriate specimens, "Judge Cooper's cabinet, being now the best adapted for the purpose, of any in the United States, Colonel Gibb's [*sic*] excepted. . . . The collection of between three and four thousand specimens, consists of his own collection; of the late Rev. Mr. Melsheimer's, and of M. Godon's." [44]

Unfortunately, these ambitious plans were doomed to early frustation. Dependent on student fees for remuneration and finding these inadequate, Cooper also conducted a private school in which Samuel George Morton, Augustus Jessup, and other talented Philadelphians were trained and encouraged in chemistry, mineralogy, and geology. But even these extra exertions were not sufficient to sustain a livelihood. In 1820 Cooper accepted an appointment in chemistry and mineralogy at the University of South Carolina after Jefferson's efforts to make a place for him at the University of Virginia had been thwarted by religious groups in that state. A few years later

the Faculty of Physical Science and Rural Economy was abolished at the University of Pennsylvania and the land intended for an associated botanical garden was sold.

Thus far our account of the development of mineralogy in Philadelphia has focused almost entirely on three institutions, the American Philosophical Society, the Chemical Society of Philadelphia, and the University of Pennsylvania. Before turning to the progress of mineralogy in the natural history societies of Philadelphia, it will be well to summarize what had been accomplished in the scientific study of minerals under the aegis of natural philosophy, especially chemistry.

Through the efforts of Woodhouse, Seybert, Hare, Cooper, and a few others the physical and chemical analysis of American minerals had begun, though on a limited scale since mineralogy was considered an adjunct of chemistry, which in turn was studied chiefly in connection with medicine. Woodhouse, Seybert, and Coxe had assembled modest collections of minerals, Seybert had described some of his, and the growing collections of the American Philosophical Society had been arranged in scientific order by Silvain Godon. Apart from occasional publications like Robert Hare's *Memoir of the Supply and Application of the Blowpipe,* the main avenue of publication had been the *Philadelphia Medical Museum* and the *Transactions* of the American Philosophical Society in Philadelphia and Mitchill's *Medical Repository* and Bruce's *American Mineralogical Journal* (1810–1814) in New York. Of the mineralogical articles in the American Philosophical Society's *Transactions* the most important were Hare's account of his experiments with the oxyhydrogen blowpipe, Godon's "Observations to Serve for the Mineralogical Map of the State of Maryland," the Silliman-Kingsley memoir on the Weston meteor, and Joseph Cloud's "An Account of Experiments Made on Palladium, Found in Combination with Pure Gold."

Cloud was in charge of melting and refining operations at the United States Mint in Philadelphia. He collected minerals and experimented avidly with them, styling himself "an humble labourer in the science of analytical chemistry." [45] His discovery and experimental proof of the occurrence of palladium in combination with gold was of some importance. Palladium had first been identified by William Wollaston, who discovered it compounded with platinum. A few years later Cloud, experimenting on some odd-colored gold ingots shipped to the United States Mint from Brazil, found that they were compounds of gold and "a metal that would resist the cupel, and was soluble

---

[43] "Syllabus of the Lectures of Thomas Cooper, Esq. M. D. as Professor of Geology and Mineralogy in the University of Pennsylvania," Miscellaneous Communications to the American Philosophical Society 2: p. 66.

[44] *Analectic Mag.* 10 (1817): p. 352. See also Thomas Cooper to David B. Warden, Philadelphia, June 24, 1818, Warden Papers, Maryland Historical Society. Cooper writes that he is enclosing a syllabus "of my mineralogical, or rather geological lectures, of which I have delivered 2 courses, but I labour under difficulty for want of books & specimens. . . ." He adds that Maclure's donations to the Academy of Natural Sciences of Philadelphia have been very helpful to him. In a letter to Parker Cleaveland dated Baltimore, December 24, 1819, Robert Gilmor, Jr. says that he has had certain experiments on phosphate of iron done for him by "Mr. T. Cooper of Philadelphia (one of the best chemists)" (Cleaveland Correspondence, Bowdoin College Library).

---

[45] Joseph Cloud, "An Account of Experiments Made on Palladium, Found in Combination with Pure Gold," *Trans. Amer. Philos. Soc.,* O.S., **6**, 2 (1809): pp. 407–411.

in nitric and nitro-muriatic acids." Having isolated this metal, Cloud found it to have exactly the same properties as palladium derived from crude platinum. These experiments, wrote Archibald Bruce, editor of the *American Mineralogical Journal*,

satisfactorily confirm the existence of Palladium as a substance possessing properties which entitle it to a place among the noble metals. We are also indebted to him for having pointed out this new metal in combination with gold, without the presence of platinum.
Since the discovery of Palladium by Dr. Wollaston, other new metals have been obtained from the ore of Platinum; and we are happy to learn that Mr. Cloud has been extensively engaged in experimenting on Rhodium, one of these new metals, which he has obtained in great purity, an account of which, we hope, he will ere long lay before the public.[46]

Cloud's "An Account of some Experiments made on Crude Platinum and a New Process for separating Palladium and Rhodium from that Metal" appeared in the *Transactions* of the American Philosophical Society in 1818, although he had read the paper before that body in November, 1809. Cloud used a magnet to separate the predominantly iron gangue from the platinum ore and then submitted the latter to the action of boiling nitric and hydrochloric acids until action ceased. He next added ammonium chloride to the solution and quickly separated the precipitated ammonium chloroplatinate, heated it to ignition to drive off the ammonium chloride, redissolved it in boiling nitric and hydrochloric acids, and obtained the metallic platinum from the resulting precipitate by directing streams of oxygen and hydrogen gas on it at white heat. The metal, described as very ductile, obviously contained impurities, since he reported the specific gravity as 23.543.

He placed plates of zinc in the original solution, causing the precipitation of palladium and rhodium, and after washing and drying the precipitate, added four times its weight of fine silver, and cupelled the mixture with lead. This process, Cloud wrote, produced a compound of silver, platinum, palladium, and rhodium, with perhaps a little gold, and after reducing the product by rolling into thin plates, he placed these in a solution of boiling nitric acid until the silver and platinum dissolved. He decanted the solution and added excess hydrochloric acid to it, which caused the silver to precipitate as the chloride. He filtered the precipitate, added either potash or prussiate of mercury to the solution, and fused the resulting precipitate with borax, obtaining pure palladium, which, he reported, had a specific gravity of just less than 11.4.

Cloud then turned his attention to the precipitate which remained after the silver and palladium had been dissolved in nitric acid. He subjected it to nitric and hydrochloric acids, which upon heating dissolved

the platinum and gold and left rhodium as a fine black powder which Cloud washed and heated to a white heat, obtaining the rhodium after fusion with a blowpipe. He reported its specific gravity as 11.2, described its color as resembling cast iron, and stated that it was rigid and friable under the hammer. But he was puzzled about the peculiar chemical properties of rhodium:

It is an extraordinary fact, first discovered by Dr. Wollaston and fully confirmed by my experiments, that rhodium in an uncombined state, and in some of its combinations with other metals, is insoluble in nitric and nitro-muriatic acids. It is particularly remarkable that it should be soluble in its native combination with crude platinum, and become insoluble in the artificial compound produced. . . .[47]

Cloud postulated that natively the platinum, palladium, and rhodium are in a state of perfect combination, that the initial action of the acids destroys this combination, and thereby the rhodium is left in such a finely divided state that it also dissolves.

But mineralogy was not entirely dependent on chemists for its progress. At the same time that chemists were analyzing mineral substances in their laboratories, natural historians were busy collecting specimens for their cabinets. Charles Willson Peale had an extensive collection of minerals accumulated by donations from every part of the United States and even from Europe. Thomas Jefferson donated grit stone, isinglass, crystals, concretions, lead ore, and silver collected by Lewis and Clark on their expedition. Benjamin Franklin contributed minerals from Derbyshire, England. The Abbé Haüy sent a collection of minerals from Paris, William Maclure another from the District of Maine, Benjamin Silliman some fragments of the Weston meteor of 1807, Stephen Elliott some calcareous spar of Virginia, Samuel Latham Mitchill ferruginous sand from the falls of the Mohawk River. There were also numerous donations from ship captains and from mineralogists in Philadelphia and its vicinity. At first these minerals were arranged according to Richard Kirwan's system, but later according to the system of Professor Parker Cleaveland of Bowdoin College after the publication of his *Elementary Treatise on Mineralogy and Geology* in 1816.

How much scientific use was made of the mineral collection in Peale's Museum is uncertain, but it must have done much to stimulate public interest in mineralogy and geology and may have been used by Benjamin Smith Barton in his lectures on natural history. The only naturalist who mentioned the collection as worthy of notice was Baron Albert von Sack, who

---

[46] *Amer. Mineral. Jour.* **1** (1814) : pp. 4, 242.

[47] Joseph Cloud, "An Account of Some Experiments Made on Crude Platinum and a New Process for Separating Palladium and Rhodium from That Metal," *Trans. Amer. Philos. Soc.*, n. s. **1** (1818) : p. 163.

visited the Museum in 1807 on his way home from Surinam.[48]

Since natural philosophy predominated over natural history at the American Philosophical Society, it was inevitable that the natural historians of Philadelphia should eventually form their own society like the Linnaean Society of London, founded in 1788 by Sir James Edward Smith and his fellow naturalists. The first such attempt occurred in 1806 when Dr. Benjamin Smith Barton, professor of materia medica, natural history, and botany at the University of Pennsylvania, organized a group known first as the American Botanical Society Held at Philadelphia and soon afterwards as the Linnaean Society of Philadelphia, "exclusively devoted to the cultivation of the different branches of natural history."

Barton's interests were largely botanical, zoological, and linguistic rather than mineralogical, but he had a consuming interest in all the productions of nature in the New World and a strong desire to encourage their study in every possible way. Both his father and his older brother had been collectors of minerals. Since Peale's Museum had already assembled an impressive zoological collection, Barton proposed that the Linnaean Society concentrate its energies on minerals and plants. "Of all the branches of natural history," he observed, "there is not one which has been cultivated with so little attention, zeal, or success, in the United States, as Geology, or Mineralogy." [49] In response the Society established a Mineralogical Committee with Charles J. Wister, Samuel Hazard, James Cutbush, and Walter Channing serving as the first members, and the public was invited to submit mineral specimens for examination and analysis free of charge.

Samuel Hazard was the son of the Samuel Hazard a catalog of whose mineral collection Franklin had sent to London in 1764. James Cutbush, a chemist by trade, lectured on chemistry and mineralogy at St. John's Church in Philadelphia and contributed to Bruce's *American Mineralogical Journal* when that publication was founded in 1810. In 1811 he organized the Columbian Chemical Society as a successor to the society founded by Woodhouse. Charles J. Wister was a retired Philadelphia merchant who acquired an interest in chemistry while attending Woodhouse's lectures and subsequently became an enthusiastic collector of minerals through association with his neighbor Adam Seybert. Impressed by Seybert's cabinet of minerals, Wister began one of his own and was frequently thereafter to be seen "armed with

specimen box and hammer, scouring the country in all directions in search of minerals." By exchanges with other collectors and purchasing what he could not collect he soon built up a cabinet second only to Seybert's. Godon was also active in the Society during his first few years in Philadelphia. News of the Society's activities and researches was published in Barton's *Philadelphia Medical and Physical Journal*, which expired shortly before Bruce founded his *American Mineralogical Journal.*[50]

The Linnaean Society of Philadelphia was short-lived, but its place was soon taken by another organization dedicated to the same purposes and destined to achieve early and lasting success. Late in January, 1812, six devotees of natural history gathered at the apothecary shop of John Speakman at the northwest corner of Second and High Streets to discuss measures for establishing an academy of natural sciences. Of the six—John Speakman, Jacob Gilliams, John Shinn, Jr., Nicholas J. Parmantier, Camillus M. Mann, and Gerard Troost—only Troost had received substantial scientific training. Born in Bois-le-Duc, Holland, in 1776, Troost earned degrees in medicine and pharmacy at the universities of Leyden and Amsterdam. In 1807 Louis Napoleon, then King of Holland, sent him to Paris for further study with the Abbé Haüy, after which Troost toured Europe collecting minerals for the royal cabinet in Holland. Appointed to serve on an expedition to Java, he became a prisoner at Dunkirk when the ship on which he was traveling was captured by a privateer. Soon after his release, Troost embarked for the Far East via America. Arriving in Philadelphia in 1810, he learned that Java had been seized by British forces. He then applied for American citizenship, married, and settled down in Philadelphia, where he established a pharmaceutical and chemical laboratory and undertook the manufacture of alum at Cape Sable, Maryland.[51]

With Troost as president, the newly founded Academy of Natural Sciences of Philadelphia got off to a vigorous start, with mineralogy as one of its chief interests. Trained in the French school of crystallography, Troost had brought with him a reflecting goniometer of the kind invented by William Wol-

---

[48] The authors are indebted to Charles Coleman Sellers of Carlisle, Pennsylvania, for this information about Peale's mineral collection. Mr. Sellers is writing a history of Peale's Museum.

[49] Benjamin Smith Barton, *A Discourse on Some of the Principal Desiderata in Natural History, and on the Best Means of Promoting the Study of This Science, in the United States* (Philadelphia, 1807), pp. 51–52.

[50] One of the circulars of the Society and a list of officers and committees are printed in the *Amer. Mineral. Jour.* 1 (1814) : pp. 53–54; other references to the Society occur on pp. 30–31, 79–80, 86–88 of that journal. See also James Mease, *The Picture of Philadelphia* (Philadelphia, 1811), pp. 303–305; *Philadelphia Medical and Physical Journal.*, suppl. 3 (May, 1809) : pp. 309–311; Charles J. Wister, Jr., *The Labour of a Long Life: A Memoir of Charles J. Wister* (2 v., Germantown, Pennsylvania, 1866) 1 : pp. 92–94.

[51] "Gerard Troost," *Dictionary of American Biography* 28 : pp. 647–648; Smith, *Chemistry in Old Philadelphia*, pp. 82 ff.; S. G. Gordon, "Pioneers in American Mineralogy," *Pennsylvania Arts and Sciences* 2 (1937) : pp. 165–168, 195. See also Gordon's account of mineralogy in early Philadelphia in the *Amer. Mineralogist* 4 (1919) : pp. 16–17.

laston, a collection of crystals, and a thorough knowledge of the latest methods in mineralogy. Within two months of its organization the Academy organized an excursion to the mines of Perkiomen Creek "to procure some minerals as the nucleus of a collection," [52] with Silvain Godon and Zacchaeus Collins, a Philadelphia merchant interested in botany and mineralogy, as guides. Soon afterwards Godon's collection was sold at auction, presumably to satisfy his creditors, but Benjamin Smith Barton outbid the Academy. Adam Seybert then offered his collection, containing nearly 2,000 specimens, to the Academy for $750.

The opportunity to acquire these treasures was too good to lose. After the collection had been examined and approved by Gerard Troost and John Shinn, Jr., the members of the Academy were given a chance to buy shares of interest-bearing stock sufficient to provide funds for the purchase of the cabinet. This stock was to be redeemed by the Academy as soon as its financial condition permitted. In July, 1813, the curators of the Academy were authorized to procure "five hundred large & fifty small paper boxes for containing the minerals." In October a committee was appointed to procure glass cases for the exhibition of the specimens. By April, 1814, the cases had been purchased and the minerals were presumably on display.

It would be interesting to know how many of the minerals listed in Seybert's handwritten catalog were included in the purchase. It appears that the specimens collected by Lewis and Clark were not included, for they do not appear in Seybert's later catalog dated 1825. Some of the items listed in the earlier catalog appear under different numbers in the later one, but many of them are missing. There are 157 fewer specimens in the later catalog, and the proportion of European to American specimens is much higher, perhaps because Seybert had become less active in collecting American specimens while continuing to acquire new specimens on his trips to Europe.

Starting with the nucleus provided by Seybert's cabinet, the Academy's collections increased rapidly. In 1814 a box of minerals was received from the Chemical and Physiological Society of Pittsburgh. In 1816 Godon's collection, which had eluded the Academy in 1812, was presented to the Academy by Joseph Watson, who had bought it at a sale of Benjamin Smith Barton's effects after his death. In the same year William Maclure, who was to succeed Troost as president of the Academy in the following year, began his long series of gifts of specimens with ten boxes of rocks he had collected in Europe and the West Indies. In January, 1817, the curators reported that the Academy possessed "four to five thousand specimens of minerals native and foreign." Soon afterward the Mineralogical Committee reported that it

had procured a book to be kept on a table "in which it is proposed to enter the localities of the minerals found in America." [53]

With Troost and Maclure to support and inspire its activities, the Academy soon became a worthy rival of the older American Philosophical Society. Maclure's scientific interests were more geological than mineralogical, but he was an ardent supporter of all scientific endeavors, and his financial resources and extensive foreign contacts were of inestimable value to the Academy. Born at Ayr in Scotland in 1763, he made a fortune in commerce early in life and proceeded to expend it liberally in support of science and social reform. Geology was his favorite science, and he pursued it vigorously during his incessant travels in Europe and America. Traveling in a private coach, he crossed the Allegheny mountains no less than fifty times, stopping every few miles to collect specimens and frightening the inhabitants of remote regions by his unaccustomed behavior. When he came through New Haven in 1808 on his return from travels in the District of Maine, Benjamin Silliman went to his hotel, eager to see this fabled scientific traveler but somewhat apprehensive concerning the reception he might receive from a man known to be an "infidel." Outside the hotel were the horses that pulled Maclure's carriage, "lean and dull" from hauling rock specimens the length and breadth of the eastern United States. Inside, Maclure's servant was busy wrapping and packing specimens for the next day's journey. Maclure himself soon dispelled whatever misgivings Silliman may have had about his character and breeding.

Mr. Maclure was at that time in his meridian [Silliman later recalled]. Being a teetotaller, drinking nothing but water and requiring only a moderate quantity of the most common articles of food, his health was perfect, and his frame robust and vigorous, as his temperance was associated with much travelling and with mountain excursions on foot; his countenance had a ruddy glow, and his manners were in a high degree winning & attractive. His language was pure and elevated—his mind being imbued with the love of science he was successful in exciting similar aspirations in other and especially younger minds. [54]

Elected to the presidency of the Academy of Natural Sciences in 1817, Maclure aided the rapid growth of the institution by providing types, a used printing press, and space in his own house for the publication of a journal, by liberal donations of books and spec-

---

[52] Microfilm of the Minutes of the Academy of Natural Sciences of Philadelphia, Collection 502 ANSP.

[53] Ibid., See especially John Barnes's "Rise and Progress of the Academy of Natural Sciences of Philadelphia," a manuscript account written in 1816 and reproduced in Minutes ANSP, Reel 1, Frame 295 ff. Also W. S. W. Ruschenberger, A Notice of the Origin, Progress, and Present Condition of the Academy of Natural Sciences of Philadelphia (Philadelphia, 1852), pp. 34–35.

[54] Quoted in Fisher, Life of Benjamin Silliman 1: pp. 284–285. See also Samuel G. Morton, A Memoir of William Maclure . . . (Philadelphia, 1841).

imens, and by keeping the Academy in constant touch with scientific developments in Europe, especially in Paris, which he visited repeatedly during these years.

Troost, for his part, provided the stimulus which only a thorough knowledge of scientific mineralogy could supply. Described as "a most lovable man who, knowing the world, was yet a profound scholar without pedantry, eloquent, and a wit and humorist," he began instructing his colleagues at the Academy in mineralogy with a course of lectures inaugurated in February, 1813, and continued at intervals through 1815. His lectures described the various families of minerals and demonstrated the use of the goniometer in measuring the angles of crystals. When the Academy established its *Journal* in 1817, Troost became one of the leading contributors. By the time he left Philadelphia with Maclure and his boatload of scientists bound for New Harmony, Indiana, in 1825, he had published nine articles in the *Journal*. Of these contributions Edgar Fahs Smith has written:

. . . one is impressed with their thoroughness. His conclusions have lasted. His analyses of minerals were of a high order. He constantly resorted to crystallographic measurements, employing the best reflecting goniometers. He continually emphasized the necessity of attention to crystalline forms as exhibited by minerals and particularly by minerals of this country, "many of which have no analogies with those described by European crystallographers.[55]

Doubtless Troost sent copies of some of these articles to his old mentor the Abbé Haüy. In 1822, the year of Haüy's death, Haüy delivered an inscribed copy of his recently published *Traité de Cristallographie* to John Walsh in Paris with a request that it be forwarded to Troost, "to whom I am obliged for several ornaments of my collection." "Besides," Haüy added, "he has given proofs of his great knowledge of crystallography through the happy applications he has made of the principles of that science with respect to several new varieties of zircon."[56] In the year after Troost left Philadelphia his *Geological Survey of the Environs of Philadelphia: Performed by Order of the Philadelphia Society for Promoting Agriculture* (1826), was published, accompanied by a colored geological map of the region he had surveyed.

Troost's faithful adherence to Haüy's ideas is evident from his attention to the minute details of crystal forms and his proclivity to give each form a French term after the manner of his teacher. He described two varieties of strontium sulfate from Lake Erie as *trapézienne* and *épointée*, complaining that Cleaveland's description of them was vague, "being applicable as well to the varieties of sulphate of barytes, as to

those of the sulphate of strontian."[57] Similarly, Troost announced the discovery of a new crystalline form of quartz, which he called *annulaire*, explaining that it was a variety formed by the decrement of one row of molecules parallel to the summit of the rhombohedral primitive form.[58] He also described two new varieties of phosphate of lime, one termed *unitaire*, and the other *unitaire* compressed.[59] Troost contributed descriptions of many other new varieties of minerals to the *Journal of the Academy of Natural Sciences*: the *unibinaire* variety of chrysoberyl from Saratoga,[60] the *progressive* variety of andalusite,[61] and the *mixunitaire* apophyllite, which had been found for the first time in North America at Point Marmoaze on Lake Superior.[62]

He made occasional mistakes, however. For example, he gave erroneous crystallographic descriptions of willemite and zincite specimens from Franklin, New Jersey, although the disparity between his results and those of Haüy with willemite puzzled him. Nevertheless, Troost was the most knowledgeable crystallographer in the United States at that time.

Troost had commenced a mineralogical collection in 1811, and over the years had arranged it in conformity with Parker Cleaveland's system of classification, describing the crystalline forms in the terminology of Haüy. Years later, in 1874, Troost's heirs sold the collection, numbering some 6,000 specimens, to the Public Library of Louisville, Kentucky, for the sum of $18,000 plus an additional fee of $2,500 to the agents of the heirs. The specimens were received by the Polytechnic Society of Kentucky on behalf of the Library for incorporation into a Troost collection in 1882. The collection is presently located in the Museum of the Louisville Free Public Library.[63]

Troost and Maclure were but two of a large number of Academy members interested in geology and mineralogy. The most prominent among the older generation, many of whom had been members of the chemical societies founded by Woodhouse and Cutbush and of Barton's Linnaean Society of Philadelphia, were Zacchaeus Collins, Charles J. Wister,

[55] Smith, *Chemistry in Old Philadelphia*, p. 84.
[56] René Just Haüy to John Walsh, Paris, May 3, 1822, Archives of the American Philosophical Society.

[57] Gerard Troost, "Description of Some Crystals of Sulphate of Strontian from Lake Erie," *Jour. Acad. Nat. Sciences Philadelphia* 2 (1821) : pp. 300–302.
[58] Gerard Troost, "Description of a New Crystalline Form of Quartz," *ibid.* 2 (1821) : pp. 212–214.
[59] Gerard Troost, "Description of Some New Crystalline Forms of the Minerals of the United States," *ibid.* 2 : pp. 55–58.
[60] Gerard Troost, "Description of a New Crystalline Form of the Chrysoberyl," *ibid.* 3 (1823) : pp. 293–295.
[61] Gerard Troost, "Description of a New Crystalline Form of the Andalusite," *ibid.* 4 (1824) : pp. 122–123.
[62] Gerard Troost, "Description of a New Crystalline Form of Apophyllite, Laumonite, and Amphibole, and of a Variety of Pearlstone," *ibid.* 5 (1825–1827) : pp. 51–56.
[63] The authors are indebted to Professor Clifford Frondel of Harvard University for this information concerning Troost's collection.

and Solomon W. Conrad, each of whom had his own cabinet of minerals. Wister, as previously noted, was a good friend of Adam Seybert and an avid collector. Collins, a merchant, is best known today for his botanical interests, but Wister considered him "the first mineralogist in our part of the country," and Silliman and Gibbs thought him worthy to be elected one of the eight vice-presidents of the American Geological Society when it was formed in 1819. Solomon W. Conrad, father of the paleontologist Timothy Conrad, was a printer and bookseller whose love of botany and mineralogy led him to neglect his business. His house, we are told, was a popular meeting place for the scientists of Philadelphia, constituting "the first natural history *salon*" in that city. The range and depth of his mineralogical studies is shown by his "Table of the Constituent Parts of Earthy Minerals," published as a broadside in 1815.[64]

The table listed approximately one hundred and fifty minerals (and rocks) in alphabetical order, the majority silicates. Alongside the names, which were chiefly borrowed from Werner, were a number of columns designating the percentage, if any, of silex, alumine, magnesia, lime, oxide of iron, oxide of manganese, potash, and water, which the mineral contained. In two additional columns, Conrad indicated the name of the chemist who had made the analysis, any loss noted in the analysis, and the presence of miscellaneous ingredients.

Conrad had studiously surveyed the existing literature. The names of over twenty European chemists appear in the citations, those of Klaproth and Vauquelin most frequently. On the other hand, it is difficult to understand his purpose in publishing the table. Quartz and eight varieties of the mineral—amethyst, cat's eye, chalcedony, chrysoprase, flint, heliotrope, prase, and rock crystal—are enumerated without any indication that they are varieties. The table was of value only from the point of view of chemical analysis, since no external characters nor crystallographic properties were delineated. It may be that Conrad believed that Maclure was devoting too much attention to external characters and Troost overemphasizing crystallography and that he wished to redress the balance in favor of chemistry.

Younger generation Academy members devoted to mineralogy included Isaac Lea, Lardner Vanuxem, William Keating, J. Lukens, John P. Wetherill, Benjamin Say (brother of the entomologist Thomas Say), Augustus Jessup, Samuel G. Morton, Thomas

Nuttall, and Henry Seybert. Lea and Vanuxem, both sons of Philadelphia merchants, discovered a mutual interest in collecting rocks and minerals early in life, pursuing their hobby together passionately but with little scientific guidance until they became acquainted with Adam Seybert's collection. In 1815 both were elected to the Academy of Natural Sciences of Philadelphia. Soon afterward Lardner Vanuxem went to Paris to study chemistry, mineralogy, and geology at the École des Mines, following in the footsteps of Adam Seybert and Gerard Troost. He was joined there by a fellow-Philadelphian, William H. Keating, whose father, Baron John Keating, had served in the Irish Brigade of the French Army in the West Indies before the Revolutionary War.

One can imagine these two young Philadelphians arriving in Paris armed with letters of introduction from Troost and Maclure and inspired by the prospect of studying science with the great masters of French science. Nor were they the last to come from Philadelphia for that purpose. In 1819, the year in which Vanuxem graduated from the École des Mines, Adam Seybert arrived in Paris with his son Henry, who was promptly enrolled in the École des Mines to pursue the studies his father had so much enjoyed twenty-three years earlier. Two years later, in August, 1821, father and son returned to Philadelphia, bringing with them nine boxes of books, chemical apparatus, and minerals. The scientific reputation which had eluded the father was to be won by the son, or so the father hoped.

I am in a very debilitated state [Seybert wrote to Parker Cleaveland], & wait only to remit my health, to enable me to make a journey as a remedy for a shattered constitution. . . . My son accompanied me to Europe— he was a pupil in the School of Mines, during two years, & I hope he will pursue his chemical and mineralogical pursuits, with ardour, as soon as we can fix for that purpose.[65]

On their return to Philadelphia, Vanuxem, Lea, and the Seyberts found the Academy much better established than it had been when they left for Europe. In 1815 one of the founding members, Jacob Gilliams, had constructed a building for the Academy on a vacant lot behind his father's house on Arch Street

[64] *Cyclopedia of New Jersey Biography* (6 v., New York, 1923) 1: pp. 376–377. A copy of Conrad's "Table," "Compiled & Published by Solomon W. Conrad, No. 87, Market-St. Philadelphia, 1815," may be seen at the Library of the American Philosophical Society. See also William Jay Youmans's biographical sketch of Conrad's son Timothy in *Pioneers of Science in America. Sketches of Their Lives and Scientific Work* (New York, 1896), pp. 385–393, especially pp. 385–387.

[65] Adam Seybert to Parker Cleaveland, Philadelphia, August 17, 1821. Cleaveland Correspondence, Division of Manuscripts, New York Public Library. There are biographical sketches of Lea and Vanuxem in Youmans, *Pioneers of Science in America*, pp. 260–269, 270–278. On Wetherill see Stephen N. Winslow, *Biographies of Successful Philadelphia Merchants* (Philadelphia, 1864), pp. 137–140. Smith, *Chemistry in Old Philadelphia*, pp. 85 ff. has a brief account of Keating, who took Cooper's place at the University of Pennsylvania in 1821. The American Philosophical Society has a copy of his *Syllabus of a Course of Mineralogy and Chemistry, As Applied to Agriculture and the Arts*, 8 pp., dated September 25, 1822. Benjamin Say is discussed briefly in Harry B. Weiss and Grace Ziegler, *Thomas Say Early American Naturalist* (Springfield, Ill. and Baltimore, Md., 1931), chap. 10.

between Front and Second Streets. In 1817 the Academy had obtained a charter from the state of Pennsylvania and had launched its *Journal*. Meanwhile the members had been busy ransacking the hills and valleys of eastern Pennsylvania and western New Jersey for mineral specimens. Two localities yielded rich rewards. One was the lead mine on Perkiomen Creek, begun in 1809 but eventually abandoned because of the difficulty of working the ore by the usual methods. According to John P. Wetherill, whose family founded the first white lead works in Philadelphia, the ore contained seventy-five per cent of lead with a trace of silver, but "no plan has yet been tried by which it can be reduced with facility." At the same time, he added, "no mine in the United States had produced so great a variety of minerals as this." The minerals found there by Wetherill, Isaac Lea, Benjamin Say, Solomon Conrad, and others were described, along with other minerals of the Philadelphia vicinity, in a memoir by Lea in the Academy's *Journal* in 1818, and Wetherill gave a more detailed account of their crystal forms in 1825. Wetherill's account included the sulfide, carbonate, sulfate, phosphate, and molybdate of lead; native copper and its oxide, sulfide, and carbonate; the sulfide and oxide of iron, the sulfide and carbonate of zinc, the sulfate of barium, as well as quartz and anthracite.[66]

The other rich mineral locality, destined to surpass the Perkiomen mine as a source of minerals, was the crystalline limestone region of Franklin Furnace and Sterling Hill in Sussex County in northern New Jersey, "the foremost mineral locality in the world."[67] By the middle of the twentieth century more than 140 minerals had been found in that district, thirty-two of which were first found there and thirty of which were unique to that area. Early mining ventures there failed, largely because the zinc ore was mistaken for copper ore and the franklinite for magnetite. It was Dr. Archibald Bruce, editor of the *American Mineralogical Journal*, who first identified the zinc ore in 1810, calling it zincite. In that same year the Franklin mine and the furnace associated with it were acquired by Samuel Fowler, a medical man with scientific as well as business interests who welcomed the mineralogists of Philadelphia and New York to his property and sought their advice. Bruce and Dr. John Torrey from New York and Vanuxem, Keating, Seybert, Nuttall, Troost, and Maclure from Philadelphia were among his visitors.

Unfortunately, the rivalry to discover and describe the minerals from these and other localities produced disagreements and confusion which sometimes generated personal animosities. In 1821 Vanuxem and Keating published three papers in the *Journal* of the Academy of Natural Sciences reporting their findings at Franklin. In one they announced the presence there of automalite, the *spinelle zincifère* of Haüy, a green mineral, usually translucent and sometimes transparent, which had as its primitive form a regular octahedron with emarginated edges. Previously it had been found only in Sweden, they wrote, noting that Thomas Nuttall had classed the mineral as gahnite in an article in the *New York Medical and Physical Journal*. Their second paper described the geology and mineralogy of Franklin and listed twenty-nine minerals found there, among which they mentioned dysluite, a new spinel. It was their third paper, however, that provoked a reaction from Troost. Describing a new mineral, a silicate of calcium, manganese, and iron, which they named *jeffersonite*, they reported its primitive form as a rhomboidal prism, its hardness as intermediate between fluorspar and apatite, its specific gravity as 3.51–3.55, and its color as dark olive green. This mineral, they announced, was a new species in the Mohs genus augite spar, following immediately after the pyramidoprismatic augite spar (the pyroxene of Haüy) and distinguished from it by the absence of magnesia.

Troost could not agree. In 1823 he reviewed the the varieties of pyroxene found in America by various collectors, including Vanuxem, Keating, Wetherill, Jessup, and Say, and demonstrated that the jeffersonite of Vanuxem and Keating was really a variety of pyroxene. Keating had mistaken the base of the primitive form of the crystal for one of its sides, said Troost. The absence of magnesia was no proof that the substance was not a pyroxene, he observed, "as is seen in an analysis performed by Roux of a crystallized variety of Pyroxene, from Arendal, and in another by Klaproth of a slaggy variety of Pyroxene from Sicily." Conscious of his superior mineralogical knowledge, Troost tried to soften the blow for the two younger men. What was really needed, he said, was a good crystallographic treatise in English. In the following year, Vanuxem and Keating acknowl-

[66] John P. Wetherill, "Observations on the Geology, Mineralogy, &c. of the Perkiomen Lead Mine, in Pennsylvania," *Jour. Acad. Nat. Sciences Philadelphia* 5 (1825–1827) : pp. 305–316; Isaac Lea, "An Account of the Minerals at Present Known to Exist in the Vicinity of Philadelphia," *Jour. Acad. Nat. Sciences Philadelphia* 1, 2 (1818) : pp. 462–482. For developments at the Academy, see Ruschenberger, *A Notice of the Origin, Progress, and Present Condition of the Academy of Natural Sciences of Philadelphia*, pp. 46–66. On page 35 Ruschenberger says: "Among the first donors of minerals were Dr. Troost, Mr. Isaac Lea, Dr. Hays, and Mr. S. Hazard. Mr. Wm. Maclure presented large and valuable collections; and Mr. H. Seybert, Joseph P. Smith, and Dr. Thomas M'Euen, contributed largely to this department. The contributors to the mineralogical cabinet since its commencement have been very numerous."

[67] Clifford Frondel, *The Minerals of Franklin and Sterling Hill. A Check List* (New York, London, Sydney, Toronto, 1972), p. 1. The "Introduction" contains valuable historical notes. See also Charles Palache, *The Minerals of Franklin and Sterling Hill Sussex County, New Jersey* (United States Geological Survey Professional Paper 180, Washington, D.C., 1935), pp. 14–15.

edged that Troost was correct, saying that they had not had any crystals of the mineral, their specimens being of the lamellar variety. "It was from the solid given by the lamellar fracture of the mineral," they wrote, "and from the absence of magnesia, which earth was regarded in some measure as an essential component of pyroxene, that we were induced to consider it as new." In the same paper, however, Vanuxem went on to give the first correct description and analysis of willemite, a silicious oxide of zinc.[68]

No such final harmony emerged from the bitter dispute over Thomas Nuttall's paper on two minerals which he called yenite and chondrodite from Sparta, New Jersey, read before the Academy on April 3, 1821, and referred to a committee consisting of Troost, Keating, and Franklin Bache. Nuttall was an excellent naturalist but no chemist, and the committee's report was highly critical of his performance, both mineralogically and chemically. Irritated by this rebuff, Nuttall published a modified version of his paper in the *New York Medical and Physical Journal* and subsequently in the *American Journal of Science*. About this time Henry Seybert attacked Nuttall for claiming to have been the first to discover fluoric acid in chondrodite, though Nuttall had made no such claim. The controversy thus begun so infuriated Nuttall that he felt ready, as he wrote John Torrey, "to anathematize natural science and banish it from my recollection!"

If I am to be abused in the Academy Journal I shall send nothing more to it and throw up my subscription. . . . Three of the *Keating* faction are now united to write me down and damn everything I do. They are a set of dastards, and I shall treat them as I did Raffy [C. S. Rafinesque] with silent contempt.[69]

Henry Seybert's talents as a mineralogist were more considerable than his altercation with Nuttall might suggest. Within three years of his return to Philadelphia he was elected to the American Philosophical

Society and the Academy of Natural Sciences and contributed solid mineralogical essays to the publications of both societies. "The merit of these contributions," wrote Edgar Fahs Smith, "lies not in the presentation of new and important facts but rather in the skill exhibited by the analyst. Evidently Henry Seybert was extremely conscientious in his work. He repeated his analyses. A single result was not sufficient for him. . . ." [70]

But Seybert's heart was not in mineralogical science. In the autumn of 1824 he and his father embarked again for Europe in hope of improving the elder Seybert's health. They reached Paris after a miserable voyage, but Adam Seybert was much too sick to undertake any of the projects he had planned before leaving home. "All my journies to the German localities, for minerals, have been done away," he wrote his business agent, Charles Graff. By the spring of 1825 his thoughts were only of returning home, if possible with William Maclure, who intended to sail for America on the first of June. It was not to be. Adam Seybert died in Paris on May 2, 1825, and was buried in Père La Chaise cemetery.

With his father's death Henry Seybert came into possession of a fortune valued at more than a quarter of a million dollars. His zeal for science, which had probably been inspired more by respect for his father than by the natural bent of his mind, diminished rapidly. After some hesitation he abandoned science for philanthropy. His mineral collection, scientific books, and chemical apparatus he gave to the Academy of Natural Sciences of Philadelphia. Years later, as a token of filial piety and a symbol of the new direction his interests had taken, he endowed the Adam Seybert Chair of Moral and Intellectual Philosophy at the University of Pennsylvania

upon the condition that the incumbent of said Chair, either individually, or in conjunction with a commission of the University Faculty, shall make a thorough and impartial investigation of all systems of morals, religion, or philosophy which assume to represent the Truth, and particularly of modern Spiritualism.

His father would probably have preferred a chair in chemistry and mineralogy.[71]

From the foregoing circumstances and from the catalog of Adam Seybert's collection dated 1825 it appears that the elder Seybert continued to take an interest in the collection he had sold to the Academy in 1812 and that he prepared a catalog of the collection shortly before his death. The prefatory note at the beginning of the catalog states that the collection had been augmented by 100 specimens purchased at

---

[68] Lardner Vanuxem and William H. Keating, "Account of the Jeffersonite, a New Mineral Discovered at the Franklin Iron Works, near Sparta, in New Jersey," *Jour. Acad. Nat. Science Philadelphia* 2, 2 (1822) : pp. 194–204; Gerard Troost, "Account of the Pyroxene of the United States, and Descriptions of Some Varieties of Its Crystalline Forms," *ibid.* 3, 1 (1823) : pp. 105–124; Vanuxem and Keating, "Observations Upon Some of the Minerals Discovered at Franklin, Sussex County, New Jersey," *ibid.* 4, 1 (1824) : pp. 3–11. See also Clifford Frondel and Jun Ito, "Zincian Aegirine-Augite and Jeffersonite from Franklin, New Jersey," *Amer. Mineralogist* 51 (1966) : pp. 1406–1413. In a note to the authors Professor Frondel observes: "The pyroxenes of these localities vary widely in composition, and the exact composition of the original material will not be known until someone finds the type specimen—if it still exists, which is very doubtful."

[69] Thomas Nuttall to John Torrey, April 12, 1823, as quoted in Jeannette Graustein, *Thomas Nuttall Naturalist. Explorations in America 1808–1841* (Cambridge, Mass., 1967), pp. 165–166. For a fuller account of this controversy see below, chap. 3.

[70] Smith, *Chemistry in Old Philadelphia*, p. 48.

[71] See Helen S. Remer, *Adam and Maria Sarah Seybert Institution for Poor Boys and Girls; Fifty Years of Service 1907–1957* (Philadelphia, 1957), pp. 3–4; Smith, *Chemistry in Old Philadelphia*, pp. 51–52.

Freiberg in Saxony in 1820 and cataloged separately according to the Wernerian system instead of being classified, like the other minerals in the collection, by the system adopted in Parker Cleaveland's *Elementary Treatise on Mineralogy and Geology,* first published in 1816. This catalog contains 157 fewer entries than the earlier one. Some, though by no means all, of the American specimens listed in the earlier catalog appear under different numbers in the later one. There are also new American entries, including a considerable number of specimens from Franklin Furnace and Sparta, New Jersey. Items 989 and 996 are listed as having been "analized [*sic*] by H. L. Seybert." [72] In general, however, the collection was now heavily weighted in favor of European specimens. The fate of the specimens from the Lewis and Clark expedition listed in the earlier catalog is not known. Fortunately for the historian of mineralogy, the collection in its final form remains intact in its mahogany chest of drawers at the Academy of Natural Sciences, each specimen tagged with a number corresponding to an entry in the catalog dated 1825.

The death of Adam Seybert and the departure of Maclure, Troost, and others for New Harmony in 1825 marked the end of the second period of development of mineralogy in Philadelphia. During the first period, which ended with the appointment of Coxe to the chemical chair, the imprisonment of Godon, and Seybert's abandonment of science for politics, mineralogy was cultivated as an adjunct of chemistry, which, in turn, depended heavily on its connection with medicine for support, although practical applications in pharmacy, chemical manufacture, brewing, assaying and the like were important in the careers of men like Seybert, Hare, Cutbush, and Joseph Cloud. Institutionally, the American Philosophical Society played the leading role, seconded by the Chemical Society of Philadelphia and the Linnaean Society of Philadelphia.

The founding of the Academy of Natural Sciences in 1812 gave mineralogy and the earth sciences generally a strong new impetus. Led by Troost and Maclure, whose professional competence, enthusiasm, and

close contacts with the French mineralogists provided inspiration and support for the younger generation of Philadelphia scientists, the Academy added rapidly to the original nucleus provided by the collections of Seybert and Godon and emerged as an important center of publication on American mineralogy through its *Journal.* After many vicissitudes mineralogy had found a home in Philadelphia and a solid base for future development.

IV. THE NEW YORK MINERALOGISTS

By the end of the eighteenth century New York was gaining rapidly on Philadelphia in population and commerce, but she still lagged behind her sister city in scientific reputation. In 1767, only two years after the founding of the School of Medicine at the College of Philadelphia, King's College had acquired a medical faculty as distinguished as the Philadelphia group. After the Revolution, however, while Benjamin Rush, James Woodhouse, Benjamin Smith Barton and their colleagues were making Philadelphia the medical center of the new nation, the New York physicians were quarreling among themselves, founding rival schools and rival journals.

The efforts of Franklin, Rittenhouse, and the Vaughans resuscitated and reinvigorated the American Philosophical Society, but the literati of New York were unable to form a literary and philosophical society until 1814, and that was defunct by 1830. Not until the founding of the Lyceum of Natural History in 1817 did New York acquire a scientific society that was to endure and make solid contributions to science. Until that time the nearest thing to a scientific society in the state of New York was the Society for the Promotion of Agriculture, Arts, and Manufactures, founded in 1791 and incorporated in 1792. In 1804 it was reincorporated as the Society for the Promotion of Useful Arts. In 1798 the Society followed the state government to Albany, where it was provided with a suite of rooms in the Capitol.

The aims of this society were severely practical. "Our business generally is to collect all the improvements within our reach that may be made in agriculture, in manufactures and in the arts of whatsoever kind they may be," declared Simeon De Witt in his address to the Society in 1799.[1] Although the roster

---

[72] "Catalogue of Minerals in the Adam Seybert Collection 1825," Collection 141, IIg, Library of the Academy of Natural Sciences of Philadelphia. A prefatory note reads:
"The object of this List is especially to preserve the localities of the specimens composing the collection. There was not the least idea of a minute description of them when the list was made out; that work is reserved for a future period.
"The rocks, consisting of one hundred specimens, which are numbered with black ink on a *white* ground, compose a collection independent of the general one—it was purchased at Freyberg, in 1820; it is arranged according to the Wernerian system & the numbers correspond with those in the German manuscript catalogue, which accompanies them."
On the page opposite the first entries is written in a different hand: "Arranged according to Cleaveland's Mineralogy."

[1] Simeon De Witt, "Address delivered before the Society, in the Assembly Chamber in the City Hall, in the city of Albany, February 20, 1799, by Simeon De Witt, Surveyor-General," *Trans. Soc. Promotion of Agriculture, Arts, and Manufactures* 1, 4 (1799) : p. 2. That the Society was not without some pretensions to science is shown by De Witt's apologetic statement on page 17:
"This state it must be confessed does not unfold to the philosophical world a much admired feature in the portrait of America. By what institutions for the furtherance of useful knowledge, or by what extraordinary efforts of our citizens to advance the arts of peace, truly patriotic, have we distin-

of members included potentates from both political parties, most of the officers and active members were Jeffersonian Republicans dedicated to agriculture as a way of life. Led by their president, Robert R. Livingston, they contributed articles on manures, sheep husbandry, grasses and hedges, bot worms in horses, and similar subjects, conceding a secondary importance to manufactures and the mechanic arts but warning that "these, though the contrary has often been believed and taught, are not the *mistresses,* but the HANDMAIDS of Agriculture." [2]

By 1815, however, the members' zeal for agriculture had declined to the point that the annual orator was led to observe that

although the improvement of agriculture was a paramount object in the formation of the Society, it is becoming a matter of secondary consideration among the members who usually attend the meetings. The encouragement of manufactures, the improvement of the mechanic arts, and the perfection of those arts depending on chemical processes, seem to afford the ordinary subjects of discussion and to engage our principal attention.[3]

In any case, the members were still agreed that "practical utility, more than recondite science, is the leading object of our society." [4]

---

guished ourselves? A university, colleges and academies we have it is true; they serve to save us from the shame of not following the fashion of nations. Nothing extraordinary can be placed to our credit on their account. What more have we to boast of? How few of our learned characters step aside for a moment from their professional employments, to show themselves interested in diffusing a taste for useful philosophy and works of ingenuity. . . .

"The society of which I now stand the representative, is without question the most consequential in the State: Which besides it, receives any notice from abroad, or is calculated to excite it? Barren as our printed transactions may appear to the prejudiced eyes of those who have not condescended to compare them with others, I will venture to affirm that they have as good a complexion, and are fully as interesting as those of a similar kind, by which Europeans are climbing up to greatness, and ascending the ladder of philosophic fame." It is worth noting, however, that Benjamin De Witt chose to publish his "Account of Some of the Mineral Productions of the State of New York (Accompanying Specimens transmitted for the Cabinet of the American Academy of Arts and Sciences)" in the *Memoirs* of the American Academy rather than in the *Trans. Soc. Promotion of Agriculture, Arts, and Manufactures.* See *Memoirs Amer. Acad. Arts and Sciences* 2, 2 (1799) : pp. 73–81.

[2] Samuel L. Mitchill, "Address to the Agricultural Society and both Houses of the Legislature of the State of New York, at Their Annual Meeting in the Court House, in the city of Albany, February 7, 1798," *Trans. Soc. Promotion of Agriculture, Arts, and Manufactures* 1, 3 (1798) : p. xii.

[3] Thomas C. Brownell, "Annual Address, on the Theory of Agriculture; Delivered by Appointment, before the Society for the Promotion of Useful Arts, at the Capitol, in the City of Albany, on the 14th February, 1815," *ibid.* 4, 1 (1816) : p. 14.

[4] Jacob Green, "An Address on the Botany of the United States, Delivered Before the Society for the Promotion of Useful Arts, at the Capitol, in the City of Albany, on the 9th day of February, 1814," *ibid.* 3 (1814) : p. 72.

Despite the utilitarianism of the New Yorkers, perhaps because of it, mineralogy gained an early foothold in New York City, especially among the medical men. Through the influence of Drs. Samuel Latham Mitchill, David Hosack, Archibald Bruce, and others, the Society for the Promotion of Agriculture, Arts, and Manufactures was alerted to the advantages that might accrue to agriculture and industry from a careful survey of the mineralogical resources of the state. The leader in this movement was Samuel Latham Mitchill, physician, professor of chemistry, natural history, and materia medica, editor, state legislator, congressman and promoter *par excellence* of useful knowledge, described variously by his contemporaries as "more of a natural philosopher than a physician," "a delphic oracle . . . in natural philosophy," "a successful promoter of physical science," "the delight of a meeting of naturalists," "the pioneer investigator of geological science among us," "the father of natural history in this State," "the Congressional dictionary," "a chaos of knowledge," and "the Nestor of American science."

In the prime of his manhood [wrote his younger contemporary John W. Francis], Dr. Mitchill was about five feet ten inches in height, of a comely, rather slender and erect form; in after life he grew more muscular and corpulent, and lost somewhat of that activity which had characterized his earlier days. He possessed an intelligent expression of countenance, an acquiline nose, a gray eye, and full features. His dress at the period he entered into public life was after the fashion of the day, the costume of the times of the Napoleonic consulate; blue coat, buff-colored vest, smalls, and shoes with buckles. He was less attentive to style of dress in his maturer years, and abandoned powder and his cue. . . . His robustness preserved his full features, and to the last not a wrinkle ever marked his face. . . .[5]

An indefatigable worker possessed of an exceptional memory, an inveterate hiker, a friend of fisherman and soap boilers as well as statesmen and scientists, an ardent patriot and organizer of civic projects, known both at home and abroad as the editor of the *Medical Repository,* Mitchill was admirably equipped by temperament, training, and profession to take the lead in organizing the pursuit of science in a city more famous for devotion to practical pursuits than for love of learning.

Born and reared on a Long Island farm, Mitchill seems to have acquired his interest in mineralogy during his medical studies in Edinburgh in the 1780's. According to James Hall, he returned from Europe with "the first mineral collection brought to this country." This collection formed the nucleus of a natural history cabinet that would grow rapidly in

---

[5] John W. Francis, *Reminiscences of Samuel Latham Mitchill, M.D., LL.D.* (New York, 1859), p. 28. For a biography of Mitchill see Courtney R. Hall, *A Scientist in the Early Republic, Samuel Latham Mitchill 1764–1831* (New York, 1934).

the ensuing years and eventually take its place in the collection of the New York Lyceum of Natural History, organized by Mitchill in 1817.[6]

Soon after his return from abroad Mitchill joined with others in founding the Society for the Promotion of Agriculture, Arts, and Manufactures. Probably at his own suggestion, he was commissioned by the Society to make a mineralogical and geological survey of eastern New York, from Long Island to the Catskills. According to Mitchill, one of the main purposes of this survey was to determine whether there were deposits of coal located within carting distance of one or more river landings. His negative finding on this point must have been disappointing to his colleagues in the Society and to the members of the New York legislature who listened to his address in Albany in February, 1798, but Mitchill himself was exhilarated by his discovery of "some new and important facts . . . which throw great light upon the history and theory of the earth."[7] He had spent five or six weeks examining geological formations from the northern and southern shores of Long Island Sound to the summits of the Catskill Mountains, clambering over the rocks along the coast, studying the rocky cliffs along the west bank of the Hudson ("climbing aloft, or walking below, or examining them with a good spying glass, as he sailed slowly by them"), tracing carefully the termination of the granitic formations and the beginning of the schist near New Cornwall, noting the strike and dip of these and the limestone formations in the Highlands, collecting fossils in the latter, and ascending the sandstone ranges of the "Blue Mountains" behind Kingston. The outcome was a highly useful report, submitted to the Society in October, 1796. Published in part in the Society's *Transactions* and in full in the *Medical Repository*, it was the first systematic geological survey of a substantial region of the United States by an American.

Although it was entitled "A Sketch of the Mineralogical and Geological History of the State of New York," Mitchill's report was more geological than mineralogical in modern terms. For the purposes of "agricultural description" Mitchill divided the region he had surveyed as follows:

I. Continent—subdivided into:
  1. The *granite country*, extending from the Sound to the termination of the Highlands, or first range of mountains.
  2. The *slate country*, beginning where the granite ends, and underlaying [sic] all other strata, fur-

ther westward, and northward, than the researches of the Commissioner have extended.
  3. The *limestone country*, spreading in some places to considerable extent, and though scattered in various parts of the country, in large bodies, yet always superficial, and bottomed upon shistic [sic] or granitical rock.
  4. The *sand-stone country*, composing the Kaats-Kill, or Blue Mountains, and some smaller strata, but always resting upon some deep-laid and more ancient fossil bodies.
  5. The *alluvial country*, consisting of horizontal layers of clay, loam, sand, turf, and generally speaking, all such matters as constitute intervale space between mountains and flat land, along creeks and rivers.

II. Islands: and these may be classed as they are,
  1. *Primaeval*, as New York, Staten-Island, and the north side of Long Island; or,
  2. Secondary, like the beaches, hassocks, and marshes on the seacoast, and all that part of Long-Island lying south of the spine or ridge of hills which runs through it from east to west.[8]

Mitchill devoted his report largely to the delineation of these formations and their relationships. On occasion, he described the mineralogical composition of the formation under discussion, using the nomenclature and classification of Swedish mineralogist Axel F. Cronstedt's *Essay Towards A System of Mineralogy*. Granitical rocks, Mitchill told his readers, "consist chiefly of quartz, feldspath, glimmer [mica] schoerl and garnet, blended together in a great variety of forms." The sandstone near Verdrietege Hook, on the other hand, "is composed of silicious grit, clay, and glimmer, with now and then a thin layer of whitish or bluish clay, or clay and mica interspersed; and sometimes bits of semi-transparent quartz . . . found in it." In the Catskill Mountains the sandstone "is grounded upon slate of a brittle and shivery texture, (*shistus fragilis*) some of which, when exposed to a high degree of heat, melts and gives evidence of a bituminous quality (*shistus pinguis*). . . . In some places, quartz (*quartzum amorphum*) is blended with the slate; and in others, veins and fissures of the shistus are filled up with it (*quartzum granulatum*)."[9]   These descriptions, interspersed

[6] S. L. Mitchill, *Catalogue of the Organic Remains and Other Geological and Mineralogical Articles, Contained in the Collection Presented to the New York Lyceum of Natural History . . .* (New York, 1826).

[7] Mitchell, "Address to the Agricultural Society . . . ," p. xxviii.

[8] Samuel L. Mitchill, "A Sketch of the Mineralogical and Geological History of the State of New York . . . ," *Medical Repository* 1 (1797–1798): pp. 293–314, 445–452; 3 (1799–1800): pp. 325–335. The first section only of this "Sketch" was published in the *Trans. Soc. Promotion of Agriculture, Arts, and Manufactures* 1, 4 (1799): pp. 124–152. The quotation is from *Medical Repository* 1: p. 284.

[9] *Ibid.*, p. 296. Axel F. Cronstedt (1722–1765), Swedish chemist and mineralogist and superintendent of mines in Sweden, discovered the element nickel in 1751. His *An Essay Towards a System of Mineralogy*, published in 1758 and translated into English in 1770, proposed a system of classification comprising four classes: (1) minerals which do not dissolve in water or oil and retain their properties with the addition of mild heat; (2) non-metallic combustibles, which dissolve in oil and burn in fire; (3) salts, which dissolve in water and yield

throughout Mitchill's report, give evidence of his careful observation.

This is particularly true of his description of the formation of peat. In the mineralogical textbooks of the time, such as Richard Kirwan's *Elements of Mineralogy*, peat was treated as a mineral substance. Mitchill's observations while he was still a medical student in Scotland convinced him that peat bogs consisted of accumulations of plant debris, especially the decayed remains of a small moss plant designated *Sphagnum palustre* by Linnaeus. Continuing his researches on this subject in America, Mitchill traced the process by which peat and turf were formed. In lowland bottoms where *Sphagnum palustre* flourished, he noted, bogs of moss were formed, providing a soil for new generations of the same species to grow upon. Since the mass of vegetable material retained water like a sponge, it became a habitat also for aquatic plants of various sorts; these in turn added to the accumulation of organic debris. This process continued as long as there was enough water to keep the moss alive, after which it soon came to an end.

From all these particulars it will be apparent [Mitchill concluded], that both turf and peat, when pure, ought to be considered merely as a residuum of decayed vegetables; that the minerals frequently found in it are foreign and casual admixtures, by no means essential to its nature, and that it ought not to be considered as a mineral production, nor classed as such in the systems; but that the clay, marl, pyrites, and other fossil bodies found among it, should be referred to their proper places in the mineralogical arrangement.[10]

In keeping with the practical purpose of his tour, Mitchill made numerous observations on how the minerals of the region were being used or might be used in the future. He noted that granitic and basaltic rocks served as materials for fences, docks, and house foundations, that slate served as roofing material, calcareous earth for making quicklime, sandstone for building, and peat for manure and fuel. Although he was unsuccessful in locating deposits of coal, he reported extensive mining of iron ore, both bog iron and mountain ore.

From among the granite rocks in the Highlands, . . . the inhabitants bring many tons down to the landings on the Hudson: thence it is carried away by water, and forged not only in the works higher up the river, but in those which are established in the Eastern states. This iron ore, some of which is magnetical, continues among

the mountains on the west side of the river, for a long distance into New Jersey, and is so abundant, that the works of Stirling, Ringwood, Mount Hope, Hibernia, Rockaway, &c. have been supplied with all their ore from these internal and neighbouring sources.[11]

Mitchill thought it best for republican virtue that no precious metals had been discovered: "Fortunately for our peace and happiness, no sources of gold and silver appear to have been detected. It is to be hoped our country contains none but those of productive labour and active industry." [12]

Soon after submitting his "Mineralogical Sketch" to the Society for the Promotion of Agriculture, Arts, and Manufactures, Mitchill organized a group known as the American Mineralogical Society to promote "the investigation of the mineral and fossil bodies which compose the fabric of the Globe; and, more especially, for the natural and chemical history of the minerals and fossils of the United States." [13] Mitchill himself was president, librarian, and cabinet keeper. Two medical colleagues and co-editors of the *Medical Repository*, Dr. Elihu H. Smith and Dr. Edward Miller, served as secretary and treasurer respectively.

The Society's aim, said Mitchill, was "to arm every hand with a hammer, and every eye with a microscope." To this end the Society circulated an advertisement soliciting information about the minerals of the United States, especially those useful in national defense, such as stones suitable for gun flint and deposits of sulphur, saltpeter, and lead ore. Apparently strained relations between the United States and both Britain and France were uppermost in the minds of the members of the Society at the time of its founding. A second advertisement, published somewhat later, took a broader view of the national benefits to be gained from exploring the mineral resources of the country.

Mineral substances enter directly or indirectly into almost every manufacture, whether of objects ornamental or useful [the Committee declared]. Glass, porcelain, gunpowder, certain of the most powerful acids, some of the most elegant and permanent of our colours and dyes, and the most powerful class of remedies known to the medical art, are chiefly of this class. . . .

A nation which is deficient either in mineral riches, or in a knowledge of them, is wanting in the most essential requisites of political and commercial independence.[14]

The advertisement concluded with a plea for the formation of mineralogical societies in various parts of the United States:

The societies formed might keep up a regular correspondence, and might send each to all the others, parts of the specimens they receive, together with the written accounts of them, whenever those specimens are of a kind

a taste; and (4) metallic minerals. Cronstedt distinguished between simple minerals and rocks composed of several minerals; he excluded fossils from his system of classification.

[10] *Ibid.*, p. 434. See the discussion of this "discovery" of Mitchill in William W. Mather's *Natural History of New York, Part IV, Geology of New York. Part I. Comprising the Geology of the First Geological District by William W. Mather* (Albany, 1843), p. 243. Mather adds: "The term mineral at the present time includes . . . peat as an organic combustible alluvion."

[11] *Ibid.*, p. 290.
[12] Mitchill, "Address," p. xxix.
[13] *Medical Repository* 1 (1797–1798) : pp. 114–115.
[14] *Ibid.* 2 (1798–1799) : p. 215.

and in sufficient quantity to be divided. It would also be useful to analyze parts of the specimens received, and to communicate the result. We conceive also, that it would not be necessary to confine our cabinets to objects merely *mineral:* whatever tends to illustrate the history of the earth and of its component parts, might, perhaps, be admitted with advantage.

A correspondence of this kind, if actively pursued a few years, would furnish our country with several valuable cabinets of mineralogy at little or no expense. These would be repositories where persons inclined to investigate such subjects, either for amusement or profit, might resort for information. They would enable the inhabitants of any one part of the union to take a view of the mineralogy of the whole United States.[15]

Despite its laudable aims and the zeal of its president, the American Mineralogical Society had only a brief career. The last notice of it as an existing organization was in the *Medical Repository* in 1801. But its general purposes were carried forward with undiminished energy by Mitchill. In the pages of the *Medical Repository* (1797–1824) and in the footnotes he appended to the American edition of William Phillips's *Elementary Introduction to the Knowledge of Mineralogy* (1818), Mitchill gave the public an account of leading discoveries, both practical and scientific, relating to the minerals of America. When gold was discovered in Cabarrus County, North Carolina, in 1803, Mitchill's devotion to republican virtue did not prevent him from acquiring specimens of the metal for his cabinet and giving a circumstantial account of the quantity and quality of the gold discovered, the methods of working it, and the total value ($11,000) of the money coined from it during the year 1804.[16] He also reported discoveries of huge deposits of lead and copper in the Louisiana Territory and as a member of Congress from 1801 to 1813 was able to describe the efforts of the federal government to obtain full information about them. Closer to home, he noted the existence of extensive marble quarries in New England, New York, Pennsylvania, and Maryland, cobalt ore in Connecticut, slate quarries in New York and Pennsylvania, iron works in New York, New Jersey, and Pennsylvania, and deposits of chromate of iron [chromite] near Baltimore and ochreous earths in New Jersey, both of which were used in producing valuable pigments. The chromate of iron, he explained, "furnishes the means of preparing the beautiful paint called the chromic yellow, with which the carriages and furniture are now painted, in Baltimore." [17] In western New York, Mitchill reported, petroleum was collected from the surface of certain waters and sold under the name of Seneka Oil: "It is employed both inwardly and outwardly, as a remedy; and is extolled by the people, for its efficacy against diseases, more especially of the stomach, intestines, kidneys, skin, joints, &c." Although he was a physician, Mitchill did not venture a professional opinion concerning the medicinal value of this cure-all.[18]

On the scientific side the *Medical Repository* served as an avenue of publication for American mineralogists and as a clearing house for mineralogical news. Besides Mitchill himself, Archibald Bruce, Thomas P. Smith, Samuel Akerly, David Bailie Warden, and others contributed brief mineralogical articles, such as Bruce's "Notices concerning Fluate of Lime and Oxide of Manganese, Discovered in the State of New York." Among the many items of mineralogical news, none excited Mitchill more than the report that Mr. Charles Hatchett of the Royal Society of London had discovered a new metal in a collection of minerals sent to Sir Hans Sloane from North America and subsequently acquired by the British Museum. Hatchett's account in the *Philosophical Transactions* of the Royal Society of London for 1802 described the specimen in question as "a dark-coloured heavy substance, which attracted my attention on account of some resemblance which it had with the Siberian chromate of iron, on which at that time I was making experiments."

Upon referring to Sir HANS SLOANE's catalogue [Hatchett continued], I found that this specimen was only described as "a very heavy black stone, with golden streaks," which proved to be yellow mica; and it appeared, that it had been sent, with various specimens of iron ores, to Sir HANS SLOANE, by Mr. WINTHROP, of Massachusets [*sic*]. The name of the mine, or place where it was found, is also noted in the catalogue; the writing however is scarcely legible: it appears to be an Indian name, (Nautneauge;) but I am informed by several American gentlemen, that many of the Indian names . . . are now totally forgotten, and European names have been adopted in the room of them. This may have been the case in the present instance; but, as the other specimens sent by Mr. WINTHROP were from the mines of Massachusetts, there is every reason to believe that the mineral substance in question came from one of them, although it may not now be easy to identify the particular mine.[19]

From his chemical experiments on the specimen Hatchett concluded that it consisted of a combination of iron with an acidifiable metal (metallic acid) different from any hitherto known to science. Although

---

[15] *Ibid.,* pp. 216–217.
[16] *Ibid.* 12 (1809) : pp. 192–193. See also William Phillips, *An Elementary Introduction to the Knowledge of Mineralogy . . . With Notes and Additions on American Articles, By Samuel L. Mitchill* (New York, 1818), p. 201.
[17] Phillips, *Mineralogy,* p. 174 n.

[18] *Ibid.,* p. 233 n.
[19] Charles Hatchett, "An Analysis of a Mineral Substance from North America, Containing a Metal Hitherto Unknown," *Philos. Trans. Roy. Soc. London* 92, 1 (1802) : pp. 49–66. The paper was read November 26, 1801. See also "Medical and Philosophical News," *Medical Repository* 6 (1803) : pp. 212, 322–324. In 1949 this element was officially named niobium.

he was unable to obtain the pure metal from its oxide, he was sufficiently certain of its novel character to propose a name for it: *columbium*. "My researches into the properties of this metal," he added,

have . . . been much limited by the smallness of the quantity which I had to operate upon; but I flatter myself that more of the ore may soon be procured from the Massachuset [*sic*] mines, particularly as a gentleman now in England, (Mr. SMITH, Secretary to the American Philosophical Society), has obligingly offered his assistance on this occasion.

Thomas P. Smith, as we have seen, was killed by a gun explosion on his return voyage to the United States, but Samuel Latham Mitchill reported Hatchett's discovery to American readers and determined the original locality of the ore containing columbium [now called niobium]. After interviewing Francis B. Winthrop of New York he concluded that it was at the head of the harbor in New London, Connecticut (formerly called Nautneague) and that the "Mr. Winthrop" who had sent the specimen to England was John Winthrop II, first Governor of Connecticut and a charter member of the Royal Society of London.

Since the time of Hatchett and Mitchill, debate has continued over both the element columbium itself and the derivation of the type specimen. In 1809 William Wollaston attempted to prove that the metal was identical with tantalum, discovered in Sweden in 1802. About 1840 Heinrich Rose reached the conclusion that columbite contained an element different from tantalum and called it niobium. Two decades later Jean Charles Galissard de Marignac concluded from extensive experiments on a variety of allied minerals that columbium and tantalum were distinct elements and that Rose and others had been led into error by the presence of titanium in all tantalites and columbites, though he was unable to obtain titanium from any of these minerals. In 1905 Edgar Fahs Smith found himself in the same position after exhaustive experimentation:

How to free the columbium from titanic acid we do not know. We are in precisely the same position as that of Marignac, notwithstanding we have probably made greater efforts than he to remove it from the columbium.[20]

The origin of Hatchett's specimen has proved equally controversial. It may have been included in the gift of approximately three hundred rocks and minerals sent to the Royal Society by John Winthrop III, grandson of the Governor of Connecticut, in 1734, but specimen No. 2029 in Sir Hans Sloane's catalog, the specimen used by Hatchett, does not correspond to any item on the list of Winthrop's donation. After detailed investigation Raymond P. Stearns, the latest

scholar to examine the problem, leaves the question open:

It is entirely possible, of course, that Winthrop [III] presented to Sloane or to the Royal Society other specimens for which no record has been found other than that given in Sloane's catalogue. It is also possible that specimens from the Royal Society's Repository found their way into Sloane's collections, including items presented in the 1660's by Winthrop's grandfather, the Governor of Connecticut. Indeed, the Sloane designation of the columbite as "from Nautneauge. From Mr. Winthrop" . . . may argue that the columbite derived from the earlier Winthrop's gifts to the Royal Society—especially as the place of origin was given as "Nautneauge."[21]

Mitchill's contributions to mineralogy were broader than his activities as organizer, investigator, author, and editor. He was also a teacher. In 1792, at the instigation of the Society for the Promotion of Agriculture, Arts, and Manufactures, Columbia College established a professorship designed to provide lectures on natural history, chemistry, and agriculture, and appointed Mitchill the first incumbent. According to Mitchill's own account, his course, conducted upon the "new French system," included "not only the classification and arrangement of natural bodies, but also . . . a great variety of arts which form the basis of medicine, Agriculture, and other useful arts, as well as of manufactures."

Any gentleman who wishes to study Chemistry may attend this class, without regularly entering College, or performing the tasks required from students on the establishment. There is a handsome apparatus belonging to this department, and a considerable collection of fossils.[22]

The collection of "fossils" [earthy substances] was probably the one Mitchill is reported to have brought back from Europe, although the College seems to have acquired collections of its own as time went on. For several years Mitchill had more medical students than undergraduates, since the course was not required for the bachelor's degree, but in 1800 chemistry was made a part of the regular college curriculum. Apparently Mitchill's lectures continued to range over the whole

---

[20] Edgar Fahs Smith, "Observations on Columbium and Tantalum," *Proc. Amer. Philos. Soc.* **54** (1905) : p. 157.

[21] Raymond P. Stearns, "John Winthrop (1681–1747) and His Gifts to the Royal Society," *Publ. Colonial Soc. Massachusetts, XIII, Trans. 1952–1956* (Boston, 1964), p. 231. See also Clifford Frondel, "Early Mineral Collections in New England," *Mineralogical Record* (Spring, 1970), p. 607. Frondel says: "The mineral itself was named columbite, so-called from Columbia, a name for America. The particular specimen was found on an unknown date in Connecticut, at a place called Neatneague or Naumeag by the Indians. Unfortunately, the exact site cannot now be identified; although it almost certainly came from a pegmatite in the Middletown area. A colored lithograph of the specimen is included in a publication by J. Sowerby, *Exotic Mineralogy*, printed in London in 1811." See also Jessie M. Sweet, "Sir Hans Sloane, Life and Mineral Collection," *Natural History Mag.* **5** (1935) : pp. 49–64, 97–116, especially pp. 115–116.

[22] Samuel L. Mitchill, *The Present State of Learning in the College of New York* (New York, 1794), pp. 9–10.

field of physical science, for the outline of them published in the *Medical Repository* in 1810 included sections on geology (embracing cosmogony, geognosy, mineralogical chemistry, and physical geography), photology, pyrology, hydrology, pneumatology, and mineralogy.[23]

When the College of Physicians and Surgeons was organized in 1807 in response to widespread dissatisfaction with the medical school at Columbia College, Mitchill was appointed professor of natural history and botany in the new institution. Presumably his lectures there did not stress mineralogy, since his colleague Dr. Archibald Bruce was professor of mineralogy and materia medica. A few years later, however, when Bruce was eliminated from the faculty in the factional strife accompanying the merging of the medical school of Columbia College with the College of Physicians and Surgeons, Mitchill was free to include mineralogy within the domain of natural history. According to his colleague Dr. John W. Francis:

His first course of lectures on natural history, including geology, mineralogy, zoology, ichthyology, and botany, was delivered *in extenso*, in the College of Physicians and Surgeons in 1811, before a gratified audience, who recognized in the professor a teacher of rare attainments and of singular tact in unfolding complex knowledge with analytic power. There was a wholesome natural theology, blended somewhat after the manner of Paley, with his prelections, and an abundance of patriotism, associated with every rich specimen of native mineral wealth.[24]

Another account of Mitchill's lectures in 1811, perhaps also written by Francis, describes Mitchill's views on mineralogical chemistry. The author states that Mitchill abandoned the quadruple classification of the Swedish mineralogist Torbern Bergman, grouping minerals under four different heads:

1. As they are united to oxygen constituting oxyds and acids; such as potash, soda, ammoniac, and all other calces of metals.
2. As combined with phlogiston, forming metals proper, and all other combinations with the principle of inflammability.
3. As being neither connected with oxygen nor phlogiston; as in massicot, flowers of zinc, and finery cinder.

4. As being associated both with oxygen and phlogiston at once, making such compounds as the mercurius praecipitatus per se.

When there are formed amalgams and alloys by union of metals with each other, such as glass, potter's ware, bricks, brass, pewter, coin, and all other metallic mixtures, they are all to be referred to the second class; and when, by the aid of sulphur, they form ores, it is to be understood, that if the metallic nature of brimstone shall be confirmed, then ores will also be placed among the metals proper, which, in this sense, will embrace, together with the metals in their ordinary forms, all the amalgams, alloys, and ores.[25]

By "phlogiston" Mitchill appears to have meant hydrogen, for this account of Mitchill's lectures indicates that Mitchill stressed

the importance of phlogiston (or hydrogen) in the constitution of metals, affirming anew, that a little phlogiston, combined with much metallic matter, made a metal, while much phlogistous gas, (inflammable air) and a little metal, formed a combination peculiarly favourable to ascent into the atmosphere, and, by inflammation, to furnish the materials of meteoric or atmospheric stones.[26]

A set of lecture notes surviving in the Morton Pennypacker Long Island Collection at the East Hampton Free Library provides additional information about Mitchill's lectures. From internal evidence it appears that these notes were written about 1819, but it is not clear whether the lectures were given at Columbia College or at the College of Physicians and Surgeons. It makes little difference, however, since Mitchill probably said much the same thing to both groups of students. The notes are in a handwriting resembling Mitchill's and seem to approximate a verbatim transcript of his lectures. The geological and mineralogical lectures begin at Lecture 54, following a much longer section of the course devoted to zoology and paleontology. Mitchill divided the history of the earth into cosmogony, geognosy, geography, and mineralogical chemistry. After a brief discussion of various cosmogonies and an exposition of the Wernerian nomenclature, classification, and general theory of the earth, he took up the different formations one by one, beginning with the primitive rocks, describing their mineralogical composition. His remarks were illustrated with numerous specimens from his own collection. "By bringing thus the elements of minerals before you," he explained, "you learn the geolog. [*sic*] alphabet & become able to combine them so as to understand compound rocks."

---

[23] "Outline of Professor Mitchill's Lectures on Natural History, in the College at New York, Delivered in 1809–10, previous to his departure for Albany, to take his seat in the Legislature of the State," *Medical Repository* 13 (1810) : pp. 257–267; *ibid.* 3 (1800) : p. 205 (on inclusion of chemistry in the curriculum for the B. A. at Columbia College). The *Medical Repository* 1 (1797–1798) : p. 257, reported that a Mr. Owens, recently arrived from Belfast, had given a mineral collection to the "Chemical Museum" at Columbia College. See also the *New York Medical and Physical Jour.* 1 (1822) : pp. 364–367, where Mitchill reports on various mineralogical, zoological, botanical, and other specimens received by "our collection at the College."

[24] Francis, *Reminiscences of Samuel Latham Mitchill,* pp. 15–16. A circular issued by the College in June, 1811, indicated that Mitchill's lectures were illustrated "by a large collection of Mineralogical and Geological specimens."

[25] "Cultivation of Natural History in the University of the College of New York. Communicated for the Register, by a Correspondent," *Amer. Med. and Philos. Register* 2 (1811) : p. 163. Francis was one of the editors of the *Register.* The quotation is from the second edition, published in 1814.

[26] *Ibid.,* pp. 161–162. The author adds : "It is pleasing to observe, that the doctrines on this subject, taught in our university college, more than two years ago, by Dr. Mitchill, have recently been inculcated in the royal institution of London, by Dr. [Humphry] Davy."

The specimens furnished much of the interest in the course. They included a box made of blue feldspar from Carinthia presented by Colonel Gibbs; tourmalines from Ceylon, Philadelphia, and Brittany; syenite from Syene ("exactly on the tropic of Cancer, where at the summer solstice Empedocles [actually Eratosthenes] observed the sun to shine directly into the well,"); parts of a huge mass of native iron found near the headwaters of the Red River in the Louisiana Territory ("where it was looked upon as sacred by the Indians"); a piece of the temple of Aesculapius at Herculaneum; garnet from Norway ("The rhombs being divided show that these dodecahedral figures make up altogether 24 equilateral triangles! Such is the mathematical precision of Nature's hand writing"); native magnesia from Hoboken, New Jersey ("first recognized by Dr. Bruce. . . . The discovery of this was one of the prettiest things that occur to any one. He established it & has it confirmed by such men as Vauquelin &c. This native magnesia is peculiar to this region—of it therefore we ought to be proud").[27]

The section of the course devoted to mineralogy proper was fairly brief. Mineralogy was defined as the "systematic arrangement of minerals acc. [sic] to their chemical constitution with some regard to their external configuration, but without reference to their geological natures." Two Swedish mineralogists, Torbern Bergman and Axel F. Cronstedt, were credited with having established the classification of minerals on a scientific basis, after whom further improvements had been made by Abraham Werner, Richard Kirwan, Robert Jameson, and the Abbé Haüy ("By his goniometer i.e. measure of the sides & angles of a crystal, he tells you its chemical nature!"). Among American mineralogists, Mitchill added, Dr. Archibald Bruce had won a "just celebrity" through his *American Mineralogical Journal*. The lectures closed with a discussion of Humphry Davy's experiments on the bases of "earths," which showed them to be metallic.

From the foregoing account of Mitchill's lectures it should be evident that his knowledge of mineralogy was general and discursive rather than analytical and experimental. His collection of minerals seems to have been more remarkable for its size and variety than for its systematic arrangement. He was familiar in a general way with the work of some of the leading mineralogists of his time, but he does not seem to have studied their works closely or to have attempted to apply their methods of research toward making new discoveries. Speaking of the magnesium hydroxide found in New Jersey, he wrote:

I had this mineral in my possession for several years, without knowing what it was. Nor could I learn from any of my mineralogical friends what opinion ought to be formed of it. But we judged by external characters only. Dr. Bruce at length undertook a chemical analysis of the mineral; and proved it to be a native magnesia. . . .[28]

It was Bruce, not Mitchill, who ascertained the nature of the red oxide of zinc, although Mitchill had received a specimen of the same substance several years earlier from someone in New Jersey. "He wished me to make experiments upon it; but I did not find opportunity or inclination to analize [sic] it. I however gave specimens freely to my mineralogical friends: and among others, to Dr. Bruce." [29] Although he taught chemistry in Columbia College and included a section on mineralogical chemistry in his lectures on natural history, Mitchill seems to have attempted relatively little in the chemical analysis of minerals. Perhaps this is not surprising in view of the range of his interests and activities. Mineralogical analysis was not easy, and very few early American chemists were skilled in it. Parker Cleaveland at Bowdoin College appealed to Benjamin Silliman at Yale for assistance in this respect, and Silliman himself only gradually gained competence in the art.

While Mitchill never became an expert mineralogist, his contributions to the development of mineralogy and geology in his native state were notable. Through the Society for the Promotion of Agriculture, Arts, and Manufactures, the American Mineralogical Society, the Literary and Philosophical Society of New York, and the Lyceum of Natural History of New York, through the *Medical Repository* and his annotated editions of Phillips's *Mineralogy* and Cuvier's *Essay on the Theory of the Earth*, through his service in Congress and the New York state legislature, and through his lectures at Columbia College and the College of Physicians and Surgeons, he called attention to the importance of discovering and exploiting the mineral resources of the United States and of cultivating the sciences for practical reasons as well as for their own sake. Many of the next generation of New York scientists owed their enthusiasm for mineralogy and geology to his teaching and example. When Samuel Akerly, Mitchill's brother-in-law, published his *Essay on the Geology of the Hudson River, and*

---

[27] Manuscript notes of lectures by Samuel L. Mitchill, vol. III, Morton Pennypacker Long Island Collection, East Hampton (New York) Free Library. Attention was called to the existence of these notebooks by Courtney R. Hall's *A Scientist in the Early Republic*. Hall (p. 149) speculates that these lectures were probably given in 1809, but internal evidence proves that they were given considerably later. In Lecture 59 it is stated that "Dr. Mitchill informed the class & showed them a new work on Geology & Mineralogy by Prof. Cleveland [sic] of Bowdoin College Maine." Cleaveland's *Elementary Treatise on Mineralogy and Geology* first appeared in 1816. Again, Lecture 51 begins: "By way of news gent. I inform you that Maj. Long of the Engineer Corps has transmitted to me a box of Louisiana minerals." Long's first expedition took place in 1819–1820. The authors are very grateful to the Trustees of the East Hampton Free Library for granting permission to examine these notebooks.

[28] Phillips, *Mineralogy*, p. 117 n.
[29] *Ibid.*, pp. 221–222 n.

the *Adjacent Regions* in 1820, he wrote it as a letter to Mitchill, dedicating to him the appended map "as a testimonial of respect by his friend and former pupil":

I am indebted to you for the first direction given to my mind in turning it to the contemplation of the works of creation [Akerly wrote]. . . . I was first led to the subject of mineralogy, by a question of yours to me while yet a school-boy on Long-Island. You picked up a stone, and inquired what it was; to which I answered, a stone. "True," you replied; "and so is this a stone," (picking up another,) "but it is evident at sight that there is a great difference between them: in what, then, does this difference consist?" This led to further observations on the appearance of rocks and stones. These observations excited my curiosity, and produced reflection on the subject; a cabinet was commenced, and I began to study mineralogy, of the fruits of which I beg you to participate.[30]

With the establishment of the Lyceum of Natural History in 1817, with Mitchill as founder and first president, Mitchill's contribution to the development of natural history in New York was brought to a fitting climax. It was with good reason that James Hall, in his report on the geology of the Fourth District of New York, called Mitchill "the father of natural history in this state."

Second only to Mitchill as a promoter of mineralogy in New York and throughout the United States was Dr. Archibald Bruce, Mitchill's colleague at the College of Physicians and Surgeons in the years 1807–1811. Born in New York in 1777, Bruce was the son of Dr. William Bruce, chief medical officer in the British army in North America. Despite his father's responsible position in the forces opposing American independence, Bruce was reared and educated in New York. In 1791, after finishing his preparatory studies at a boarding school on Long Island, he entered Columbia College, where, against his father's wishes, he began to think seriously of a medical career. At the same time he began to acquire a liking for mineralogy through the influence of Dr. David Hosack, who had taken him on as a private medical pupil. Hosack had just returned from his medical studies abroad, bringing with him a collection of minerals which he later presented to Princeton College. While in London, Hosack had attended a course of lectures on mineralogy by a pupil of Abraham Gottlob Werner named Johann G. Schmeisser.[31]

Hosack communicated his enthusiasm for the science to his pupil, and Bruce began his own collection of minerals. In 1798 he embarked for Europe, bent on acquiring a medical degree from the University of Edinburgh, the Mecca of American medical students since the middle of the eighteenth century. Geology and mineralogy flourished at Edinburgh. James Hutton had published his two-volume *Theory of the Earth* in 1795, and the controversy between the Huttonians and the disciples of Abraham Gottlob Werner was the issue of the day at the University.

After earning his M.D. in 1800, Bruce set out on a tour of Europe, seeking out the leading mineralogists in each city he visited and offering them American minerals from his own collection in return for European specimens. In London he became acquainted with Charles Francis Greville and the Count de Bournon. An eminent mineralogist, Bournon had emigrated to England during the French Revolution. There he arranged Greville's splendid mineral collection, which was later acquired by the British Museum. In Paris, probably through the good offices of Greville and the Count de Bournon, Bruce met the Abbé Haüy and his colleagues, forming friendships that were to stand him in good stead after his return to America. Whether Bruce attended lectures on mineralogy and geology at the Museum of Natural History and the School of Mines is not known, but it seems likely that he, like Adam Seybert a year or two earlier, took advantage of the opportunity to hear public lectures by the celebrated scientists of Paris.

From Paris he went on to Switzerland and Italy, augmenting his collections as he went. Here again the details of his travels are missing, but we know that he visited Heinrich Struve at Lausanne and learned that another American, Colonel George Gibbs of Newport, Rhode Island, had recently been Struve's pupil in mineralogy. Returning to London, Bruce brought his five-year sojourn in Europe to a joyous climax by marrying in London. One can imagine the feelings of happiness and satisfaction that filled Bruce's mind as he returned to New York in the fall of 1803, bringing with him his bride, his medical degree from Edinburgh, and a handsome collection of minerals containing a wide variety of European specimens, including "the chief part of the new metals and other late discoveries of the chemists."[32]

Bruce devoted the first few years after his return to establishing himself in medical practice and aiding his old mentor Dr. Nicholas Romayne in setting up state and county medical societies and a college of physicians and surgeons. Meanwhile his zeal for mineralogy and his collection of minerals ("freely shown to the curious") continued to grow. "The ruling passion of Dr. Bruce's mind," his biographer later wrote

---

[30] Samuel Akerly, *An Essay on the Geology of the Hudson River, and the Adjacent Regions; Illustrated by a Geological Section of the Country, from the Neighbourhood of Sandy-Hook, in New Jersey, Northward, Through the Highlands in New York, Towards the Catskill Mountains; Addressed to Dr. Samuel L. Mitchill, President of the New York Lyceum of Natural History* (New York, 1820), pp. 3–4.

[31] "Domestic Intelligence," *Amer. Jour. Science* **4** (1822): p. 398. This notice quotes a letter from Professor Jacob Green to the editor stating that Hosack had given a collection of a thousand minerals to Princeton College.

[32] *Medical Repository* **8** (1805): pp. 433–434.

of him, "was a love of natural science, and especially of mineralogy."

Returned to his own country, after being so long familiar with the fine collections . . . in various countries in Europe, Dr. Bruce manifested a strong desire to aid in bringing to light the neglected mineral treasures of the United States. He soon became a focus of information on these subjects. Specimens were sent to him from many and distant parts of the country, both as donations and for his opinion respecting their nature. In relation to mineralogy he conversed, and he corresponded extensively, both with Europe and America; he performed mineralogical tours; he kindly sought out and encouraged the young mineralogists of his own country, and often expressed a wish to see a journal of American mineralogy upon the plan of that of the School of Mines at Paris.[33]

Recognition of Bruce's proficiency as a mineralogist soon followed. With the founding of the College of Physicians and Surgeons in 1807, Bruce was appointed professor of mineralogy and materia medica. Somewhat later he became registrar of the College. He was now in a position to continue his mineralogical studies as part of his official duties and to make his laboratory in New York a headquarters for research on American mineralogy. In these studies he was ably assisted by William Langstaff, who carried out many analyses for Bruce.

The circle of Bruce's mineralogical friends and informants grew rapidly. In New York he could count on assistance from Mitchill, David Hosack, Samuel Akerly, the chemist John Griscom, and many others. Not far away in New Haven was Benjamin Silliman, himself newly returned from scientific studies at Edinburgh and working diligently to build a mineral collection and introduce the study of mineralogy and geology at Yale College. At Newport in Rhode Island was Colonel George Gibbs, the young American whom Bruce had heard about when he visited Struve in Lausanne. Already famous for the huge and varied mineral collection he had acquired during his European travels, Gibbs moved freely between Newport, Boston, and New York, meeting mineralogists and geologists wherever he went and joining wholeheartedly in their schemes for promoting the study and exploitation of American minerals. Like Bruce, Gibbs was a friend of the Count de Bournon, through whom the two Americans had first met.

In Baltimore there was Robert Gilmor, Jr., no less enthusiastic a mineralogist than Bruce, Gibbs, and Silliman and, like them, recently returned from Europe with a collection of minerals. Gilmor's travels for business or health brought him frequently to New York, where he and Gibbs and Bruce soon became fast friends. In Philadelphia Bruce's connections were less intimate, but Zacchaeus Collins, Adam Seybert, Solomon Conrad, and others sent him specimens and welcomed him on his occasional visits. Beyond

Philadelphia, Bruce's circle of personal acquaintance stopped, but his correspondence reached all the way to Georgia, Kentucky, Ohio, and Brunswick, Maine, where Parker Cleaveland, who had fallen in love with mineralogy more or less by accident, was beginning work on the first American textbook of that science.

Abroad Bruce kept up his contacts with the Count de Bournon, the Abbé Haüy, and others and sought to expand his circle of correspondents through David Bailie Warden, Mitchill's correspondent in Paris. Warden continued his role as intermediary between American and French science long after he had lost his position in the American diplomatic service.[34] In 1808 the *Medical Repository* noted that Bruce's cabinet had been enriched with specimens of newly discovered European minerals sent by Haüy, including native carbonate of magnesia, silicious borate of lime from Arendal in Norway, and laumonite from Brittany. Both Haüy and Alexandre Brongniart made Bruce their chief avenue of contact with American mineralogists.[35]

Scattered letters provide glimpses of Bruce's contacts with his mineralogical colleagues at home and abroad. On July 10, 1807, Bruce acknowledged receipt of a letter from Gibbs, calling himself "doubly obliged to my friend Count Bournon . . . for making me acquainted with a gentleman of whom I have heard much both at home and abroad." Apparently Bournon had urged Gibbs to get in touch with Bruce on his return to America.

It has been to me a source of regret to think that in this extensive & interesting unexplored country so little attention has been hitherto paid to Mineralogy [Bruce wrote]. I trust however you will be the means of giving a spur to enquiry and that we shall have reason to acknowledge our obligations to you for the attention you have paid and the trouble you have taken to introduce among us a taste for this truly interesting & beautiful science.

When at Lauzanne [*sic*] I was pleased to hear from your old preceptor M. Struve that he had had an American as his pupil.

I shall be happy to communicate any information in my power respecting the mineralogy of our vicinity and to send you such specimens as it affords.

I am at present engaged in preparing a course of lectures which it is my intention to deliver during the ensuing winter having lately had the honor of being appointed Professor of Mineralogy by the Regents of the University of this State.

I am gratefully obliged to you for the pamphlets of M. Buch. You would confer on me a particular favor by giving me any information which may have lately trans[pired?] in Europe relative to Mineralogy.

Should you write shortly to Count Bournon or Mr. Greville I pray you to remember me kindly.[36]

[33] "Biographical Notice of Dr. Bruce," *Amer. Jour. Science* 1 (1818–1819): p. 303.

[34] See Francis C. Haber, *David Bailie Warden. A Bibliographical Sketch of America's Cultural Ambassador in France, 1804–1845* (Institut Français de Washington, 1954).
[35] *Medical Repository* 11 (1808): p. 443.
[36] Bruce to Gibbs, New York, July 10, 1807, Gibbs Scientific Correspondence, Newport Historical Society.

In September of the same year Bruce received a letter from the Abbé Haüy asking him to deliver a box of minerals Haüy was sending to Gibbs in return for a splendid specimen of granite covered with green tourmalines, some of which contained violet tourmalines.

I have noted this remarkable association in a memoir which I published some years ago on the tourmalines of the United States [Haüy wrote]. His [Gibbs's] box contains a beautiful piece of Wavellite of Devonshire, which was sent to me several days ago by M. Wavell himself, the discoverer of this mineral, some pieces of blue crystallized carbonate of copper [azurite] which were found near Lyons in France and which are very much sought after by foreigners, a large crystal of sulphate of barytes from Auvergne, and some other objects which I hope will be acceptable to M. Gibbs. I assume that M. De Kay has delivered to you my two memoirs on aragonite and on precious stones. I add here four copies of the first and three of the second, which I beg you, Sir, to be so good as to direct to their destination. These memoirs will appear in the annals of the Museum, and I am sorry not to have had more copies printed. I have also received two boxes of which one was sent by M. Mitchell and the other by M. Shaeffer [Rev. F. C. Schaeffer]. I intend to send each of them something when my collection of duplicates, which is presently almost exhausted, has been replenished by means of some small collections I am expecting from various countries. I will write them at the same time to give explanations concerning some of the interesting objects they have sent me.[37]

In April, 1808, Bruce established contact with David Bailie Warden in Paris through the good offices of Dr. Edward Miller, co-editor of Mitchill's *Medical Repository*. Warden, who was already acting as a go-between in mineralogical matters for Mitchill, Benjamin Smith Barton, Charles Willson Peale, Silvain Godon, William Maclure, and others, was equally useful to Bruce, delivering letters and specimens to Bruce's correspondents in Paris, forwarding their returns and the latest mineralogical works, and recommending Bruce to the Abbé Correa da Serra and other European scientists leaving Paris for America.[38]

Contacts between Bruce and his new-found friend Colonel Gibbs were frequent and cordial. Acknowledging a letter from Gibbs in July, 1809, Bruce wrote:

I returned after a weeks absence having traversed some of the most interesting parts of ye mining district of New Jersey. I found several new things, & among them a purple fluate of lime [fluorite] in two situations. I also obtained the Titan Silico calcaire [titanite], the Menacha-

nite [ilmenite], a yellow sulphur of zinc [sphalerite], an oxyd of manganese, & a yellow substance in a primitive carbonate of lime with the nature of which I am not altogether acquainted. Of these I send you specimens, in the Box which contains ye Journal de physique &c. which I hope you will receive safe. I shall send them tomorrow by one of the Newport packets. I hope you will favor me with your memoir on the coal formations as soon as you can conveniently. . . . You will oblige me greatly if you will have the goodness of the receipt of this to send me the spec. of the Titan you promised me as also the platina crucible as I find it very difficult to proceed without it in examining the ores of Titanium. You probably ere this have heard of the death of Mr. Greville, a circumstance greatly to be regretted particularly by those who knew him. Mr. Robert Gilmor has been spending a few days in New York as also Don Josef de Roxas of Mexico, both keen mineralogists. Do me the favor to write me soon. . . .[39]

Gilmor came to New York from time to time on business or on his way to the Ballston Spa in summertime and nearly always looked for Bruce and Gibbs. Having missed both of them in the summer of 1812, he urged Gibbs to visit him again in Baltimore with his wife and to bring Bruce along. In the following summer he wrote that he would be passing through New York on his way to New Haven and hoped that Gibbs and Bruce would accompany him into Connecticut.[40]

Bruce was also in constant touch with Benjamin Silliman, whose brother, Gold Silliman, had a business in New York. In June, 1810, Bruce wrote Silliman about the conduct of his nephew, who was a student at Yale, and invited the Sillimans to visit in New York. A year later he repeated the invitation and asked about several boxes of minerals sent to New York by "our friend Gibbs"—"I presume they are intended for Yale College." Soon afterwards Bruce recommended Dr. Samuel Akerly, "a practitioner in this City & brother in law of Dr. Mitchills, he has paid much attention to the study of Natural History, particularly Mineralogy. It will afford him great satisfaction in having an opportunity of viewing your splendid collection." Gilmor, too, visited New Haven with a letter of introduction from Bruce, but Silliman

[37] Abbé René Haüy to Archibald Bruce, Paris, September 8, 1807. The letter quoted is Gibbs's copy of the original. This copy is in the Gibbs Scientific Correspondence at the Newport Historical Society. See Haüy's "Observations sur les Tourmalines, Particulièrement sur celles qui se Trouvent dans les États-Unis," *Journal des Mines* 37 (1815), pp. 399–408; an English translation of this memoir by P. S. Townsend was published in the *Amer. Monthly Mag. and Crit. Rev.* 1 (1817): p. 127 ff.

[38] Edward Miller to D. B. Warden, New York, April 3, 1808; Archibald Bruce to D. B. Warden, New York, October 20, 1812, Warden Papers, Maryland Historical Society.

[39] Bruce to Gibbs, New York, July 14, 1809. Gibbs Scientific Correspondence, Newport Historical Society.

[40] Robert Gilmor, Jr. to George Gibbs, Baltimore, October 15, 1812; Gilmor to Gibbs, Baltimore, June 11, 1813; Gilmor to Gibbs, July 9, 1813, Gibbs Scientific Correspondence, Newport Historical Society. See also Bruce to Gilmor, New York, June 20, 1814, Dreer Collection (Physicians), Volume I, Historical Society of Pennsylvania: ". . . accept my best thanks for the Box of Minerals which arrived safe at last. They are all remarkably well defined & make a conspicuous figure in my American Cabinet which is now very nearly arranged. They are extremely interesting & tend to confirm ye opinion I had formed of ye riches of ye vicinity of Baltimore. What a wonderful similarity there is to ye Swiss Alps particularly St. Gothard & Chamounix. One specimen I observe of Granite which contains a green substance of an emerald tint which I take to be a green oxide of uranium. . . . As soon as I receive the Tin Box I shall return it with some thing new."

was away. Concerning mineral collections Bruce wrote:

I am happy to find you intend forming an American cabinet. Mine has made very considerable progress. . . . I send you by my nephew two varieties of phosphate of lime—No. 1 crystalline embedded in sulphuret of Iron. The crystals are well defined and present a variety of colors. No. 2 may be considered as amorphous. Both varieties are found in the highlands N York. I wish you would send me a general list of minerals in which you are deficient so that I may not be troubling you with duplicates. Send it soon & I will attend to it immediately, by selecting & forwarding such as I may have. . . . have the goodness to inform me if there are any opportunities for England or Sweden direct.[41]

In Philadelphia, too, Bruce had mineralogical friends and correspondents. In 1810 he heard from Adam Seybert that a Dutch mineralogist, Gerard Troost, had arrived there bearing letters for Bruce from the Abbé Haüy. In a letter to Zacchaeus Collins he wrote:

I duly received your favor my dear Sir mentioning the specimens you wished for your Cabinet, to which I have to inform you I have paid attention having selected from my duplicates such as I hope will prove acceptable tho I could have wished they were somewhat larger. As you informed me you were anxious to obtain crystalized minerals I have put up for you a small collection of the gems as Spinelle, Emerald, Hyacinth, Topaze Garnet with Idocrase Meionite Pyroxene &c. &c. The best mode of exhibiting these single crystals which I have preferred is placing them on small pedestals of wax in paper boxes as they can be more conveniently examined without danger of losing them.
I have added the metals you wished & also some of the newly discovered minerals which I have lately received from Messrs. Haüy & Bournon as the Cryolite Datholite Spudomene &c.
This collection you would have received ere this but that I found difficulty in meeting with a private opportunity. Our friend Mr. [Solomon W.] Conrad is to pass thro this [city?] shortly on his return to Phila. To him I shall give the box.[42]

As his collections and his circle of mineralogical friends and correspondents grew, Bruce decided to undertake a long contemplated project. Ever since his return from Europe he had dreamed of establishing a journal similar to the *Journal des Mines,* the well-known publication of the École des Mines, edited by the men he had met in Paris. By 1809 he was sufficiently pleased with the progress of his researches, his lectures at the College of Physicians and Surgeons, and his contacts with mineralogists at home and abroad to issue a "Prospectus of a Periodical Work To Be Entitled, *The American Mineralogical Journal,* Conducted by Archibald Bruce, M.D. Professor of

Mineralogy in the University of the State of New-York." The journal, he announced, would "elucidate the mineralogy of the United States, than which there is no part of the habitable globe which presents to the mineralogist a richer or more extensive field for investigation."

Like Mitchill, Bruce held out the prospect of important benefits to science, industry, and agriculture from the systematic study of mineralogy and geology. Americans were already making "laudable exertions" to develop manufacturing. What could be more helpful in this enterprise than researches into the nature and location of the minerals on which so many of the useful arts depended? Patriotism, no less than a desire for profit and a sincere interest in the progress of science, should lead the citizens of America to support the new journal by sending the editor communications relating to

the geology and mineralogy of particular districts; the history of mines, their products, methods of reduction, and improvements in metallurgy generally; descriptions of individual specimens, their constituent principles, localities, and uses to which they may be applied in the arts; mineral waters, their situation, analysis, and use in the arts, and in the cure of diseases.

For his part, the editor would see to it that his readers were kept abreast of the latest developments in mineralogical science and practice at home and abroad.

The United States, whose scientific journals up to this time had been restricted to the transactions of two or three scientific societies and a handful of medical journals, could now boast a periodical devoted exclusively to mineralogy and geology and their applications. And this in a country where, a few years earlier, Benjamin Silliman had been forced to take the box of minerals belonging to Yale College to Philadelphia to find someone who could identify them for him! It was a bold undertaking, perhaps a rash one, but Bruce knew that he could count on cooperation from Mitchill, Griscom, Akerly, Silliman, Gibbs, Gilmor, and the Philadelphia mineralogists, and he hoped for additional subscribers and contributors among his circle of correspondents extending from the District of Maine to Georgia and Kentucky. Contributions from abroad were not expected, but Warden, Haüy, Brongniart, and others in Paris would keep him supplied with European intelligence and do their best to advertise the new journal among the scientists of Europe.

The first issue of *The American Mineralogical Journal* was announced for January, 1810, but it did not appear until April. As might be expected, the articles were supplied largely by Bruce himself and his associates in and around New York. Samuel L. Mitchill led off with a descriptive catalog of some mineral specimens he had collected on a tour to

---

[41] The letters quoted are in the Silliman Correspondence in the Beinecke Library at Yale University: Bruce to Silliman, New York, June 13, 1810; July 25, 1811; August 7, 1811; August 13, 1811; August 16, 1811; August 25, 1813.

[42] Archibald Bruce to Zacchaeus Collins, New York, July 6, 1811, Collection No. 129, Academy of Natural Sciences of Philadelphia.

Niagara Falls in 1809. Samuel Akerly provided a geological account of Dutchess County, Colonel Gibbs a notice of the iron works at Franconia, New Hampshire, and of the West River mountain in Connecticut, Griscom a chemical analysis of the mineral springs at Litchfield, New York, and Bruce himself a notice of fluates of lime in America and a definitive description of a new mineral species, the magnesium hydroxide of New Jersey. George Chilton, an English chemist who had settled in New York as an instrument maker, furnished a chemical analysis of barytic spar from New Jersey, and Dr. William Meade, another immigrant from Britain, a description and analysis of a specimen of lead ore from the mines at St. Genevieve on the Mississippi River. From Philadelphia, Silvain Godon contributed a note on phosphated lime and phosphated lead from Pennsylvania, and Charles J. Wister described melanite from Pennsylvania and amber from New Jersey. Three detailed accounts of mineralogical publications followed, including William Meade's pamphlet on Rhode Island coal, Joseph Cloud's account of a gold-palladium alloy reprinted from the *Transactions* of the American Philosophical Society, and the "Geological Inquiries" issued by the Geological Society of London for the guidance of field workers. An "Appendix" reprinted William Wollaston's memoir "On the Identity of Columbium and Tantalum" from the *Philosophical Transactions* of the Royal Society of London.[43]

All things considered, the first number of *The American Mineralogical Journal* was a praiseworthy performance, highly creditable to the editor and his little band of contributors. If their publication did not equal the *Journal des Mines* in the range and quality of its articles, it nevertheless attracted favorable attention from the editors of that journal. In May, 1811, they informed the readers of the *Journal des Mines* of the appearance of Bruce's *Journal* and published in French translation the entire table of contents of the first issue. This was followed by a translation of Bruce's "Notice on the Fluates of Lime in America," with a promise of further excerpts from

the new journal in the future. The *Journal des Mines* had already translated William Wollaston's article on the identity of columbium and tantalum from Bruce's *Journal,* and further selections from the "Journal Minéralogique Americain" were soon to follow. The *Journal de Physique* and the *Annales* of the National Museum of Natural History also drew on articles in Bruce's publication.

According to the editors of the *Journal des Mines,* Bruce's publication was destined to appear quarterly, but Bruce soon discovered the impossibility of maintaining any such schedule. Still full of enthusiasm and confidence in June, 1810, he wrote to Benjamin Silliman:

I trust ere this you will have recd the 1 No. of the Mineralogical Journal. It was unavoidably delayed much longer than I wished. I hope however that I shall be enabled by the assistance of my literary friends to be much more punctual hereafter. I am about putting No. 4 [No. 2?] to press as I am anxious to get it out next month. I wish much to have your name among the contributors. . . .
Gibbs I am happy to find is working hard in the cause. I am now engaged in a paper (for my next number) on the ore of titanium found in the United States. You will oblige me greatly by any hints on the subject. I hope you intend with Mrs. S. to pay us a visit this summer. If so it will afford me great pleasure to have some conversation on the subject of our science.[44]

Further letters to Silliman in late August and mid-October indicated that Bruce was under pressure from his readers to produce the second number. But, although he felt "in honor bound to give it as soon as possible," it did not appear until early in 1811.

Obtaining contributions for the third number was no easy matter, as Bruce's strenuous appeal to Colonel Gibbs on July 6, 1811, indicated:

For the sake of mineralogy do send me on something for No. II [No. III?] & that too, soon, or the Journal dies. Do enlist Silliman or rather urge him to furnish somewhat. . . . I wish indeed you would write to your Min. friends. . . . I have just returned from Philadelphia. They are there rather at logger heads, somewhat envious of each other. Do anything you can for the Journal & with haste.[45]

By the end of the month, however, Bruce was in a happier mood. To Benjamin Silliman he wrote:

You no doubt have long ere this recd. the 2d No. of the Journal, which I am happy to say has greatly increased in circulation having within a very short time since received orders for its being sent to France, Germany Eng. Scotland & Ireland. No. 3 you will observe is announced for August & on Monday next it will be put to press.[46]

---

[43] The *American Mineralogical Journal* (1810–1814) has been reprinted in facsimile edition in the Hafner series "Contributions to the History of Geology," edited by George W. White. Concerning Dr. William Meade, John Torrey wrote to Amos Eaton several years later: "He is the person who is so frequently quoted in Cleaveland's Mineralogy for interesting localities. He is now [June 26, 1822] on a mineralogical tour through the Northern States, & having requested of me a list of all the mineralogists of my acquaintance, in the region he expects to visit, I gave him your name for one. . . . You will find him one of the most intelligent mineralogists you ever knew, having the history of every American mineral at *his fingers ends*. He will be of much use to you, should you wish to make exchanges of minerals." Quoted in Ethel M. McAllister, *Amos Eaton Scientist and Educator 1776–1842* (Philadelphia and London, 1941), p. 270. For information about George Chilton, see *Amer. Jour. Science* 31 (1837): pp. 421–424.

[44] Bruce to Silliman, New York, June 13, 1810. Silliman Correspondence, Beinecke Library, Yale University.
[45] Bruce to Gibbs, New York, July 6, 1811. Gibbs Scientific Correspondence, Newport Historical Society.
[46] Bruce to Silliman, New York, July 25, 1811. Silliman Correspondence, Beinecke Library, Yale University. See also Bruce's letter of August 7: "No. 3 is now in the press & will appear the latter part of this month."

Bruce's *Journal* was beginning to attract notice abroad, but troubles at home would delay the publication of the fourth and last number of the first volume for more than two years. In 1811 the faculty of the College of Physicians and Surgeons was reorganized, and Bruce was ousted from his positions as registrar and professor of mineralogy and materia medica. Bruce appears to have continued to give lectures as a faculty member in the short-lived Medical Institution of the State of New York, but the loss of his position at the College must have been a serious blow to Bruce's professional career and to his plans for promoting the study of mineralogy in New York. The controversies surrounding the reorganization of the faculty continued for some time and did much to distract Bruce from his duties as editor of *The American Mineralogical Journal.*

The fact is [Bruce wrote to Silliman in August, 1813], that the whole medical world here has been in a kind of fermentation . . . until some little subsidence has taken place I do not think it would be advisable to touch the 4th & last No of vol. 1 tho quite prepared. I am therefore compelled to keep it back a few & I hope I may add a very few days.[47]

In the same month Bruce wrote to Colonel Gibbs in a similar vein:

. . . such is the present situation of things here that 'tis impossible to say what will be done. The Journal is ready for publication & waits only something like a settled plan. As to myself my mind is made up & were it not, that by its publication which is looked to as an index of my determination I should quit it tomorrow. At all events it must appear in a few days. At least I have every reason to hope that in that time some plan will be settled. I assure you I wish most sincerely you were here. . . . I agree perfectly with you & henceforth you may say to those who you think will contribute that it certainly will appear quarterly. At least there will be four numbers yearly if as I before told you they contain but one sheet.[48]

---

[47] Bruce to Silliman, New York, August 16, 1813, Silliman Correspondence, Beinecke Library, Yale University. Kelly and Burrage, *American Medical Biographies* (Baltimore, 1920), p. 158, state that "on the reorganization of the faculty [of the College of Physicians and Surgeons] he [Bruce] and Romayne and others lectured in an extramural course." Confirmation of this statement is implied by the reviewer ("K.") of Parker Cleaveland's *Elementary Treatise on Mineralogy and Geology* in Biglow's *Amer. Monthly Mag. and Crit. Rev.* 1 (1817): p. 184: ". . . Doctors Mitchill & Bruce have, annually, for a number of years past, given separate courses on this subject [mineralogy and geology] in New York. . . . Their lectures have been thinly attended, until of late years the spirit of investigation has spread, and their hearers have become more numerous." For an account of the factional disputes which led Bruce to resign from the College of Physicians and Surgeons and join the new Medical Institution of the State of New York, see David L. Cowen, *Medical Education: The Queen's-Rutgers Experience 1792–1830* (New Brunswick, New Jersey, 1966), pp. 7–8. Bruce was Recorder in the new institution, which expired by 1817 after failing to obtain a charter from the State of New York.

[48] Bruce to Gibbs, New York, August 21, 1813, Gibbs Scientific Correspondence, Newport Historical Society.

The final plan was to publish the fourth number together with the first three in a single volume. At last, in 1814, the volume appeared, prefaced by Bruce's prospectus of the work. Comprising only two hundred and seventy pages, it represented the combined efforts of the devotees of mineralogy throughout the United States, including those like Godon, Meade, Cooper, and Chilton who had left Europe and cast their fortunes with those of the American republic. Most of the articles were original contributions appearing for the first time, and nearly all were by native or naturalized Americans. By an odd quirk of fate the author of one of the two articles reprinted from the *Philosophical Transactions* of the Royal Society of London was James Smithson, the wealthy English expatriate who, a quarter of a century later, bequeathed half a million dollars to the United States to establish an institution for the increase and diffusion of knowledge. One wonders whether Smithson recalled that one of his scientific memoirs had been reprinted in the first specialized American scientific periodical, *The American Mineralogical Journal.*

Altogether, twenty-one Americans contributed forty-two articles to the four issues of the journal. Samuel L. Mitchill and Benjamin Silliman each sent five communications, Col. George Gibbs, Dr. William Meade, and Bruce four each, and Samuel Akerly three. Most of the articles were descriptive accounts, either of minerals, of mineral waters, of the geological features of a particular locality, or of metallurgical operations. Nine were reports of chemical analyses of minerals or of chemical operations. At this stage of development American mineralogists were most interested in knowing what minerals their country contained. They included few able chemists, and none employed the crystallographic approach of Haüy and his school. Instead, they relied chiefly on Wernerian methods, transmitted to America in the works of Richard Kirwan and Robert Jameson.

Of the chemical analyses, those by Bruce on "native magnesia" [magnesium hydroxide] and "red oxide of zinc" [zincite] stand out. Bruce expressed surprise that Brongniart, in his *Mineralogy,* had described what appeared to be native magnesia from the Piedmont area as magnesite. He then went on to detail the external characters and chemical analysis of specimens of "native magnesia" found near Hoboken, New Jersey. He demonstrated the absence of lime and of carbon dioxide and other gases and concluded that the substance contained 70 per cent magnesia and 30 per cent water of crystallization. The mineral was subsequently named *brucite* in his honor.[49]

---

[49] Archibald Bruce, "On Native Magnesia from New Jersey," *Amer. Mineral. Jour.* 1 (facsimile of the 1814 edition; New York and London, Hafner, 1968): pp. 26–30. Brucite ($Mg(OH)_2$) ordinarily occurs in plates, but the Hoboken material is in part fibrous (Frondel).

Bruce's description of "red oxide of zinc" was the first account of the mineral. Again he gave close attention to its external characters and outlined his method of analysis, concluding that it contained 76 per cent zinc, 16 per cent oxygen, and 8 per cent oxides of manganese and iron.[50] A few years later a French chemist, Pierre Berthier, repeated the analysis and confirmed Bruce's findings, with the exception that his improved technique disclosed that the mineral contained only a trace of iron.[51] Discussing Bruce's analyses of these two minerals the Yale mineralogist George Brush wrote in 1882: "So thoroughly was the work done by Bruce, that these species remain today essentially as he described them, and his papers may well be studied by mineralogists now as models of accuracy and clearness of statement." [52]

Meade's analysis of galena from Missouri was less satisfactory; he reported that it contained only 72 per cent lead, whereas it normally contains about 86 per cent.[53] George Chilton reported that an accident had prevented a completely thorough analysis of a specimen of barite from Sussex County, New Jersey, but he still managed to determine the percentages of the constituents with an error of only 1 per cent.[54] Three articles on the decomposition of potash and the recovery of pure potassium reflected the interest in this subject generated by Sir Humphry Davy's experiments. The most vivid account was Benjamin Silliman's description of an explosion of silver fulminate which had nearly cost him his eyesight.[55]

The descriptive accounts of minerals vary in quality, but again those of Archibald Bruce are conspicuous. One of these detailed the external characters of fluorite that had been found in several localities, another those of various titanium ores: rutile, ilmenite, and sphene.[56] Robert Gilmor contributed an excellent description of many minerals he had detected in the Baltimore area, including apatite, gypsum, magnesite, garnet, feld-spar, tourmaline, hornblende, epidote, kyanite, and a number of metallic ores.[57]

Several articles stressed the practical applications or possible uses of minerals. William Meade, for example, reported on a deposit of low sulphur anthracite in Rhode Island and urged that it should be used in iron furnaces in order to conserve wood.[58] Bruce pointed out that zinc could be used in the manufacture of brass and of white vitriol and that the oxide of zinc could be substituted for white lead as a pigment. An article by Dr. Samuel Brown of Lexington, Kentucky, extracted from the *Transactions* of the American Philosophical Society, gave a detailed description of the Great Saltpeter Cave near Livingston, Kentucky, and emphasized its practical importance as a source of one of the main ingredients of gunpowder. John Griscom, describing the constituents of the water at Ballston Springs, inserted a note of humor when he reported that a nearby resident had consumed twenty-seven glasses of the water one morning and, finding that the water relieved him of a complication of diseases, "brought a sick horse to partake with him."

The *Journal* reprinted only three articles from foreign sources: "On the Identity of Columbium and Tantalum" by William Hyde Wollaston and "On the Composition of Zeolite" by James Smithson, both from the *Philosophical Transactions* of the Royal Society of London, and an essay "On the Topaz of Scotland" by Robert Jameson, from the *Memoirs* of the Wernerian Natural History Society. The articles by Wollaston and Smithson are of interest as portraying the contemporary quarrels and confusion generated by the phenomenon of isomorphous substitution. In his essay on zeolite Smithson chided Haüy for distinguishing natrolite as a species different from zeolite and pleaded for the retention of the name *zeolite* given to it by Cronstedt as a hydrated silicate of alumina and soda. Wollaston's article demonstrates the difficulties confronting early investigators who attempted to analyze the ores of tantalum and columbium [niobium]. The two elements are found in varying proportions, and no method had been developed to allow their separation.

It had taken much effort to produce these first four issues, but the warm reception accorded to the *Journal* both at home and abroad did much to convince Bruce that his labors had not been fruitless. In France, the editors of the *Journal des Mines* translated extensive excerpts, including Silliman's observations

[50] Archibald Bruce, "Description, and Chemical Examination of an Ore of Zinc, from New Jersey," *ibid.*, pp. 96–100.

[51] Pierre Berthier, "Analyse de deux minéraux zincifères des États Unis d'Amérique," *Annales des mines* 4 (1819) : pp. 483–494.

[52] George J. Brush, *A Sketch of the Progress of American Mineralogy: An Address Delivered Before the American Association for the Advancement of Science, at Montreal, August 23, 1882* (New Haven, Conn., n.d.), p. 10.

[53] William Meade, "Description and Analysis of an Ore of Lead from Louisiana," *Amer. Mineral. Jour.* 1 (1814) : pp. 7–10.

[54] George Chilton, "Chemical Examination of Heavy Spar from New Jersey," *Ibid.*, pp. 16–19.

[55] Benjamin Silliman, "Particulars relative to a Late Accidental Explosion of Fulminating Silver, in the Chemical Laboratory at Yale College," *Ibid.*, pp. 163–166.

[56] Archibald Bruce, "Description of some of the Combinations of Titanium, Occurring within the United States," *Ibid.*, pp. 233–243.

[57] Robert Gilmor, Jr., "A Descriptive Catalogue of Minerals Occurring in the Vicinity of Baltimore, arranged according to the Distribution Méthodique of Haüy," *ibid.*, pp. 221–232.

[58] [William Meade], "An Enquiry into the Chemical Characters and Properties of that Species of Coal Lately Discovered at Rhode-Island, with some observations on its application to the arts and manufactures of the Eastern States," *ibid.*, pp. 34–40.

on the geology of the New Haven area, Colonel Gibbs's account of West Mountain, Akerly's geological description of Dutchess County, Godon's description of Pennsylvania apatite, and Meade's account of Rhode Island coal. In the *Annales* of the Museum of Natural History, Vauquelin quoted and confirmed Bruce's description and analysis of magnesium hydroxide, and Haüy went out of his way to praise Bruce and his journal when describing specimens of cymophane [chrysoberyl] and pyroxene sent to him by Bruce.

I have just had a brochure from Haüy [Bruce wrote to Gibbs]. It is a paper on a new variety of pyroxene from the vicinity of New York. He speaks highly of the Journal & what you will say is somewhat new in an European & particularly a Frenchman, admits that they are to expect (having already recd) much information from what he calls "plusieurs savans d'un mérite distingué la plupart originaires de ce pays." [59]

Gibbs and Robert Gilmor were as delighted as Bruce with the *Journal's* reception in France. In January, 1814, Gilmor reported to Gibbs that he had just received several issues of the *Journal des Mines*, "which speak of his work [Bruce's *Journal*] as of the first importance, & the Abbé Haüy quotes it." British comment on *The American Mineralogical Journal* was relatively slight, but the President of the Geological Society of London, Horatio Greenough, wrote to William Meade that the new periodical was "particularly interesting," a comment which Meade doubtless reported to Bruce. [60]

Despite these encouraging signs that the *Journal* was valued in the highest mineralogical and geological circles, Bruce experienced increasing difficulty in providing for its continuation. In September, 1816, he acknowldeged receipt of a box of specimens from Silliman and held out the prospect that the fifth number would soon appear.

I have just recd a letter from Sir Joseph Banks (propria persona) requesting I would have the numbers forwarded to him as soon as possible *after their publication* for physicians often prescribe strong stimulants in cases in which the circulation is languid. I trust you will agree with me that such a hint from such a man as Sir Joseph will operate powerfully in producing a No. Five. I am now engaged in translating a very interesting catalogue of localities of Mexico for the Journal. [61]

But Number Five was never to appear. Bruce's health was failing, and his delight in mineralogy seems to have diminished somewhat in his last years. As time passed, Bruce's friends began to despair for the *Journal*. "Bruce is a most excellent man, and a first rate mineralogist. I fear the Journal will languish for want of being published with regularity," Gilmor wrote to Gibbs. [62] Three years later Number Five had still not been published, and Gilmor was even more impatient with Bruce. From Paris he wrote to Gibbs:

Tell my friend Dr. Bruce that he is losing his reputation with his friends here for *punctuality* (the old fault). Several complain of having sent him things and never hearing from him in return. Among them is the Count de Bournon. I shall not pardon him soon his unaccountable neglect of his promise to send me letters to his mineralogical friends. [63]

Little did Gilmor suspect the cause of his friend's neglect. Across the ocean Bruce lay dying, the victim of a stroke that would bring his life to an untimely end at the age of forty-one, just three days later. Not long afterward his collection of minerals was sold at public auction for five thousand dollars. [64]

[59] Bruce to Gibbs, New York, August 21, 1813, Gibbs Scientific Correspondence, Newport Historical Society. Haüy also paid his respects to Benjamin Smith Barton, [Charles Willson?] Peale, Silvain Godon, William Maclure, and Samuel L. Mitchill for their contributions to his collection and to David Bailie Warden for his role in promoting scientific commerce between the United States and France. See René Just Haüy, "Sur des Cristaux de Pyroxene des Environs de New-Yorck," *Annales du Muséum National d'Histoire Naturelle* 14 (1813): pp. 257–267; Haüy, "Sur les Cymophanes des États-Unis," *ibid.* 18 (1811): pp. 57–69. For other evidence of French interest in Bruce's *Journal*, including reprinting of articles translated from it, see *Journal des Mines* 29 (1811): pp. 398–399; 30 (1811): pp. 77–78, 317–320, 393–395; 33 (1813): pp. 175–186; 34 (1813): pp. 238–240; *Annales du Muséum National d'Histoire Naturelle* 10 (1813): pp. 167–168; *Journal de Physique* . . . 72 (1811): p. 219; 74 (1812): p. 470; 75 (1812): pp. 75–79; 79 (1814): pp. 215–219; 80 (1815): pp. 137–146.

[60] "Extracts from a letter written by G. B. Greenough, Pres. of Geology Soc. of London to Dr. Meade, Boston. Feb. 14, 1812, 2 Parliament Street, London," Cleaveland Correspondence, Bowdoin College Library.

[61] Bruce to Silliman, New York, Sept. (?), 1816, Silliman Correspondence, Beinecke Library, Yale University.

[62] Gilmor to Gibbs, March 20, 1815, Gibbs Scientific Correspondence, Newport Historical Society.

[63] Gilmor to Gibbs, February 19, 1818, Gibbs Scientific Correspondence, Newport Historical Society.

[64] *Amer. Monthly Mag. and Crit. Rev.* 4 (1818–1819): p. 139. The Gibbs Correspondence at the Newport Historical Society contains two letters from E. W. Bibby to Gibbs concerning Bruce's last illness and the disposal of his mineral collection. The first, dated New York, March 7, 1817, says: "I regret not having had an opportunity to converse with Dr. Bruce on the subject of the Mineralogical collection after you left us: his mind became much infeebled from a continual coma which was observed during the day and an unceasing restlessness [?] at night which precluded all possibility of transacting any business and we are left in a dilemma what to do with them. They are undoubtedly valuable so much so that his estate will not admit their gratuitous disposal. At the desire of Mrs. Bruce & the advice of his confidential friend & Attorney, as well as from my own personal knowledge of your acquaintance with the extent of the Cabinet and its value will you do me the favour to communicate your ideas on this subject and should you think probable they may be sold to advantage in any part of these United States it will be more agreable [sic] to us than sending them to Europe. Hosack has already requested the refusal of them. The price I mentioned to him was $10 or 15,000 for the whole business & European collection but of this you will better judge." Benjamin Silliman found the collection still in New York City in

Despite the expiration of *The American Mineralogical Journal* in 1814 and the death of its editor four years later, interest in mineralogy increased steadily in New York during the second decade of the century. The indefatigable Dr. Mitchill continued his efforts in behalf of the science in his lectures at the College of Physicians and Surgeons and Columbia College and the pages of the *Medical Repository*. In 1814 Colonel George Gibbs, who had moved from Newport to Boston in 1811, purchased a handsome country estate at Hallett's Cove near Hellgate, Long Island Sound, and entered actively into the scientific life of New York, sailing back and forth to the city in his yacht *Laura*. Surrounded by his books, his scientific instruments, his collections, his gardens, dogs, and horses, Gibbs soon made Sunswick a seat of hospitality for his scientific colleagues and his friends and acquaintances generally. In October, 1816, John Torrey, a young medical student who had become interested in botany, chemistry, and mineralogy through his studies with Mitchill and Hosack, wrote to his friend Amos Eaton:

I am now thick with Gibbs. He is a princely fellow. He lives 7 or 8 miles from here (on Long Island) & has a fine laboratory, well furnished. I shall go there to analyze my Datholite & his Brucite (Fluate of Magnesia & Silex). He gave me many minerals.[65]

In due time the new enthusiasm for natural history in all its branches assumed institutional form. In 1814 the literati of the city came together to organize the Literary and Philosophical Society of New York under the presidency of the Governor of New York, De Witt Clinton. The membership included Mitchill, Bruce, Hosack, Griscom, and several others interested in mineralogy and geology. As corresponding secretary, Mitchill was in touch with the Abbé Haüy, Professor Johann Friedrich Blumenbach of the University of Göttingen, James Hutton's famous disciple John Playfair, and other distinguished European scientists. At the same time Mitchill was active in securing the appointment of a committee "for

exploring this state with a view to elucidate its Natural History, Antiquities & Geography." The Committee soon recommended that efforts be made to raise three thousand dollars to pay for investigations of this kind, and a circular was issued appealing for funds to launch the project. Simultaneously Charles Edmond Genêt, who had settled in New York after his dismissal from the French diplomatic service in the famous Genêt affair of 1793, moved to establish a geological committee to collect and arrange mineralogical, geological, and paleontological specimens from the state of New York and other parts of North America. In keeping with these interests the Society heard papers on earthquakes, mineral waters, fossils, and the like and published some of these in its short-lived *Transactions*.[66]

Soon after the founding of the Literary and Philosophical Society the passion for science invaded the New-York Historical Society, which had been established in 1804, for "the collection and preservation of whatever may relate to the natural, civil, literary, and ecclesiastical history of the United States, and of this State in particular." In 1816 a cabinet of natural history was established by the members. In March of the following year the Society inaugurated lectureships in the various branches of science and appointed lecturers, who also served as chairmen of committees charged with collecting information in their respective departments. Mitchill was appointed lecturer for zoology and geology, Hosack for botany and vegetable physiology, John Griscom for chemistry and natural philosophy, and Colonel Gibbs for mineralogy. A month later a circular requesting specimens and bearing Gibbs's signature was issued by the mineralogical committee, and Gibbs reported in Biglow's *American Monthly Magazine and Critical Review* that plans were under way to display specimens from every state in the Union in specially designed cases, "one case being devoted to each state, after the manner adopted in the national collection at the École des Mines at Paris." The same magazine reported that Mr. J. B. Bogert had donated a substantial collection of minerals to the Society and that Gibbs had offered to deposit some of his ample collection in the Society's cabinet. "Our Cabinets are beginning to fill," Mitchill wrote to the secretary of the Academy of Natural Sciences of Philadelphia. "Mr. Bogert had commenced, and Col. Gibbs will follow." [67]

1859. ("Origin and Progress of Chemistry . . . at Yale College," MS at Beinecke Library, Yale University, 4: p. 25.) Bibby's second letter to Gibbs, dated New York, September 3, 1818, reads: "Anticipating daily the pleasure of seeing you I have refrained from writing: I am extremely desirous of consulting you as to the best arrangement we can fall upon in the sale of Dr. Bruces [*sic*] minerals to obtain those you particularly wish and also as to the reasonable value of the whole in case I should feel disposed to purchase myself when there will be no difficulty as to the yellow mineral. Mr. Bayard has promised the notes of our friends life as minutely as he can call them to mind and the engraving will be ready on Saturday in time I trust for the next number of the Journal." The last sentence suggests that the biographical memoir of Bruce published in Silliman's *Journal* in 1818 was written by Nicholas Bayard, or possibly by Gibbs from materials supplied by Bayard.

[65] John Torrey to Amos Eaton, New York, June 16, 1819, quoted in McAllister, *Amos Eaton*, p. 264.

---

[66] "Records of the Literary & Philosophical Society of New York," MS in the Manuscript Room of the New-York Historical Society. A further account of the Society's interest in natural history may be found in *Max Meisel, A Bibliography of American Natural History. The Pioneer Century, 1769–1865* (3 v., Brooklyn, N.Y., 1924–1929) 2: pp. 224–227.

[67] *Amer. Monthly Mag. and Crit. Rev.* 1: pp. 48, 124. Samuel L. Mitchill to Reuben Haines, Esq., New York, April 17, 1817, Coll. No. 567, Academy of Natural Sciences of Philadelphia.

In the meantime, still another organization, the Lyceum of Natural History of New York, had been formed under Mitchill's presidency. One of the vicepresidents, Rev. Frederick C. Schaeffer, was placed in charge of the mineralogical collections, and John Torrey, later to succeed Mitchill as president, was appointed one of the curators. Like the Historical Society and the Literary and Philosophical Society, the Lyceum was assigned rooms in the old Almshouse, a three-story brick building which had been set aside for literary and scientific purposes by the City of New York in 1816, when the paupers who had occupied the buildings were moved to new quarters at Bellevue. Redesignated the New York Institution of Learned and Scientific Establishments, the Almshouse now housed the American Academy of the Arts, the Literary and Philosophical Society, the New-York Historical Society, the Lyceum of Natural History, John Griscom's lectures on chemistry and natural philosophy, and John Scudder's American Museum.

Of these institutions the New-York Historical Society and the Lyceum of Natural History were to be the most enduring. It soon became apparent, moreover, that the Lyceum provided a more congenial home for the study of natural history than the Historical Society. Addressing the Society in 1820, David Hosack called on the members to consider

how far the benefit that was contemplated by the Historical Society, in the erection of this department of Natural History, will not be more essentially and completely accomplished by the Lyceum, and whether the proposed transfer of the Cabinet will not prevent collision in our labour, and by concentrating the objects in view, promote the interests of science, and advance the respectability of the New York Institution.[68]

Within a year the Society's natural history collections had been transferred to the Lyceum, which now rivaled the Academy of Natural Sciences of Philadelphia, though it lacked the financial support provided to the latter by William Maclure.

Interest in mineralogy was not confined to New York City and its environs. When the Society for the Promotion of Agriculture, Arts and Manufactures (later called the Society for the Promotion of Useful Arts) moved to Albany in 1798, it established an institutional base for mineralogical and geological investigations in the state capital. In 1810 a circular addressed to the members of the Society and to devotees of the science of mineralogy in every part of the state declared that the science had been sadly neglected in the state and that, "to do away that imputation," the Society had resolved to collect and preserve specimens of earths, fossils, and metallic ores from all parts of New York. Readers of the circular

were urged to send specimens to the recording secretary in Albany, where they would be preserved and displayed in a cabinet in the Society's rooms at the Capitol.[69]

In February, 1813, Dr. Theodric Romeyn Beck, recently graduated from his studies with Mitchill, Hosack, and perhaps Bruce at the College of Physicians and Surgeons, addressed the Society for the Promotion of Useful Arts on the mineral resources of the United States, "with their various application to the Arts, and to demonstrate the practability of the increase of different manufactures, whose materials are derived from this source." "If these pages should call the attention of any of its readers to the science of mineralogy, and cause them to lend their exertions to promote its usefulness," Beck declared in the published address, "the highest wish of the Author will be gratified." After reviewing the splendid progress of the sciences in Europe in recent decades, Beck turned his attention to the practical benefits that could be obtained from these discoveries.

With such a flood of light pouring on us from the old world, it surely becomes the duty of every one who desires his country's prosperity, to direct it to the best advantage. Our resources must be investigated, and our ability to conduct useful undertakings ascertained.[70]

To this end, Beck reviewed the various kinds of minerals, listing places where they had been found and indicating to what extent they had been mined and from what foreign countries they were imported. His report was necessarily incomplete, but he drew the encouraging conclusion that the United States was better situated than any other country in the world to make constructive use of its mineral resources.

We are united in one common aim of fostering and encouraging whatever may be useful to our country. The situation of *that* country is peculiarly favorable to such exertions. Our horizon though cloudy, is not enveloped in the lurid darkness of the Eastern hemisphere. In one state in Europe, every new invention which substitutes machinery for manual labor, inflames civil discord, while we hail it as a valuable addition to our national riches. In others, the Arts and Sciences are only cultivated that they may be accessary [*sic*] to plans of military despotism, and that contending nations may wield with fiercer destruction the infernal machinery of war. Who from this

---

[68] David Hosack, "Inaugural Address Delivered before the New-York Historical Society, on the Second Tuesday of February, 1820," *Coll. New-York Hist. Soc.* 3 (1821) : p. 279.

[69] "Domestic Intelligence," *Amer. Med. and Philos. Register* 1 (1810) : p. 122; the second edition, 1814, was used. Meisel, *Bibliography of American Natural History* 2: p. 111, says that the Society "maintained a cabinet or museum of specimens from all parts of the state, which were cared for in a room in the then new Capitol. . . . In May, 1824, it was merged with the Albany Lyceum of Natural History (1823), . . . to form the Albany Institute (1824), as its first department."

[70] Theodric Romeyn Beck, "Annual Address, Delivered by Appointment, Before the Society for the Promotion of Useful Arts, at the Capitol, in the City of Albany, on the 3d of February, 1813," *Trans. Soc. Promotion of Useful Arts, in the State of New-York* 3 (1814) : p. 11.

survey, where the exertions of human intellect appear cursed with unprofitableness, will not turn to his own country, as the last defence and shelter of civilization and human happiness? *Who* will not lend his best aid in conducting her to the summit of national greatness?[71]

Responding to Beck's exhortations, the Society for the Promotion of Useful Arts established four committees to investigate topics of interest to the Society. Among the members of the Committee on Chemistry ("which will particularly include Mineralogy, Metallurgy, Dying, Tanning, Brewing, Baking, Malting, Glass Making, Pottery, and all the Arts connected with the science of Chemistry") were Colonel Gibbs, Archibald Bruce, and two professors of chemistry at New York colleges: Thomas C. Brownell and Joseph Noyes. Bruce and Brownell were also members of another committee of the Society charged with discovering means of exploring the coal resources of the state. In its published circular the latter committee appealed to the cities and villages along the Hudson to appropriate funds for a geological survey of eastern New York.

The common council of the city of Albany have generously offered a reward of one thousand dollars to any person who may effect the wished for discovery. It is trusted that they will so alter the destination of this sum as to devote it to the promotion of the method proposed by the Society for the Promotion of Useful Arts. Such a sum from this public spirited body, together with twice or thrice its amount from the corporation of the opulent city of New York, and a moderate aid from the city of Hudson and the villages along the river, would be sufficient to defray the expense of a very thorough geological survey of the country and a sufficient number of borings to ascertain a fact no less important to the progress of manufactures than it is interesting to the economy of housekeeping.[72]

New York did not establish a state geological and natural history survey until 1836, but it is apparent that Mitchill, Bruce, and their colleagues recognized the importance of such an undertaking long before that time.

The close union of science and the practical arts advocated in Beck's address found still another champion in Amos Eaton, whose labors in behalf of American science rivaled those of Mitchill and Benjamin Silliman. During his legal training in New York City in the years 1800–1802 Eaton had studied botany, chemistry, and natural philosophy under Hosack and Mitchill. Convicted of forgery and sentenced to life imprisonment in 1811, Eaton was pardoned by Governor De Witt Clinton and aided in beginning a new career as a scientist by his old teacher Mitchill. Entering Yale College in 1816 at the age of thirty-nine, Eaton studied chemistry, geology, and mineralogy with Benjamin Silliman and carefully studied Colonel

Gibbs's mineralogical collection, which had been deposited on loan at Yale a few years earlier. "You will never be perfect in mineralogy until you are familiar with Gibbs' cabinet," Eaton wrote to his young friend John Torrey, who had visited him and learned the rudiments of botany from him while Eaton was in prison.[73]

His studies at Yale completed, Eaton gave a course of lectures on mineralogy and botany at Williams College in the spring and summer of 1817. The following year found him busily engaged in lecturing and field work in Albany and Troy, where he soon joined forces with Stephen Van Rensselaer, Theodric Romeyn Beck, Governor Clinton, and others in projects for applying science to the economic development of the state. The influence of Clinton and Beck gained him an invitation to lecture on the applications of geology and chemistry to agriculture before the state legislature in a room of the Capitol assigned for the use of the Society for the Promotion of Useful Arts. Meanwhile he was helping to organize the Troy Lyceum of Natural History.

In Albany [Eaton wrote in his *Geological Text-book*], I found Dr. T. Romeyn Beck, and in Troy, Doctors Burrett, Robbins, and Hale, zealous beyond description in the cause of Natural Science. . . . They, together with several others, had become members of the New York Lyceum of Natural History; and in the fall of 1818, established a society of the same name and upon a similar plan in Troy. Collections were made with such zeal, that in the course of a few months, Troy could boast of a more extensive collection of American geological specimens, than Yale College, or any other institution on the continent.[74]

By December, 1824, the mineralogical collection had grown to the point where Eaton could describe it as "the best collection of minerals in the *interior* of the Northern States."

That collection does not contain all the rare minerals; but it contains most excellent large fair specimens. It is contained in three large elegant cases. It is kept in the court-house, which is to be torn down next season.[75]

The energy and enthusiasm of Eaton and Beck, combined with the patronage of Stephen Van Rensselaer, soon made the Albany-Troy region a leading center of geological research. At the Rensselaer School, founded in 1824, Eaton inaugurated an imaginative program of training in pure and applied science that produced some of the leading geologists of the next era in American science. At the same time he

[71] *Ibid.*, p. 41.
[72] *Ibid.* 3: pp. 253–255, 258–259.

[73] Eaton to Torrey, New Haven, October 4, 1816, quoted in McAllister, *Amos Eaton*, p. 264.
[74] Amos Eaton, *Geological Text-book for Aiding the Study of North American Geology: Being a Systematic Arrangement of Facts, Collected by the Author and His Pupils, under the Patronage of the Hon. Stephen Van Rensselaer* (2nd ed., Albany, New York, and Troy, N.Y., 1832), p. 18.
[75] Eaton to Stephen Van Rensselaer, Dec. 14, 1825, quoted in McAllister, *Amos Eaton*, pp. 274–275.

and Beck gained Van Rensselaer's support for an ambitious geological survey extending from Boston to Lake Erie. They began with Albany and Rensselaer counties and worked their way westward. Their reports, culminating in *A Geological and Agricultural Survey of the District Adjoining the Erie Canal* (1824), which contained a profile view of the rock strata from the Atlantic Ocean to Lake Erie, paved the way not only for the Erie Canal but also for the establishment of the New York State Geological and Natural History Survey. Eaton's interests were more geological than mineralogical, but the close relationship between the two sciences insured that mineralogy developed concomitantly with geology in the regions where Eaton's influence prevailed.

The correspondence between Eaton and John Torrey reveals the cooperation mixed with rivalry that marked the relationships between the scientists in the Albany area and those in New York City. Torrey, remembered today as a leading American botanist of the nineteenth century, had a considerable reputation as a mineralogist in the early part of the century. Trained in botany, chemistry, and mineralogy at the College of Physicians and Surgeons, he arranged the mineralogical cabinet his teacher David Hosack had brought from Europe and soon became intensely interested in the chemical analysis of minerals, working in Colonel Gibbs's laboratory at Sunswick.

Eaton and his colleagues were chiefly field workers, somewhat contemptuous of geologists and mineralogists "who study hand specimens in the cabinet and are disposed to multiply names." This difference in outlook, as well as the rivalry between New York and Albany, is apparent in a controversy that broke out in 1818 over the identification of a mineral specimen which had been sent to Dr. Mitchill for the Lyceum of Natural History by the recording secretary of the Society for the Promotion of Useful Arts. The specimen was referred to the Mineralogical Committee, and in due time the Committee reported that the substance was sulphate of strontian. Mitchill included a notice of the discovery in his edition of Phillips's *Mineralogy* in 1818, stating that fibrous sulphate of strontian had been discovered west of Albany. Acting on this report, the Troy Lyceum of Natural History sent Amos Eaton, accompanied by Lt. R. C. Pomeroy of the U. S. Ordnance, to search for the mineral.

To their delight they found it in great abundance on the face of a hill near Carlisle, Schoharie County, N. Y. Soon afterward Pomeroy sent specimens of the mineral and the matrix in which it had been found to Mitchill, together with a letter, published in an account of the Lyceum's proceedings in Biglow's *American Monthly Magazine and Critical Review* in December, 1818, describing the location and geological situation of the mineral deposit. He added that a blacksmith named Elisha Baldwin had found the min-

eral to be a much better flux in brazing than borax and an excellent substitute for clay and borax in withstanding welding heat.

Meanwhile Eaton, judging solely from the specific gravity, hardness, color, and texture of the substance, had decided to accept Mitchill's identification of the mineral as sulphate of strontian. In New York, however, his friend John Torrey had analyzed a specimen and concluded that it was sulphate of barytes. This was subsequently confirmed by Benjamin Silliman at Yale and Professor Chester Dewey at Williams College, but the Albany mineralogists were not convinced. In July, 1819, Eaton sent Torrey a specimen of "Schoharieite," observing:

> Not one of us is fairly committed in relation to that mineral. E. [Edwin] James has experimented upon it with his usual accuracy. He produced the red flame distinctly from one of its salts. I certainly produced it from both muriate and nitrate. . . . With respect to spec. grav. I have found it under 3.5 and over 4.5.[76]

Torrey replied by sending Eaton some carbonate of strontian, instructing him to compare the color of its flame with that produced by the Schoharie mineral.

> If now you do not see a different colored flame from what you obtained from the mineral from Carlisle, I engage to eat a pound of it. . . . Dr. Mc Neven [professor of chemistry at the College of Physicians and Surgeons] has also just finished an analysis of the mineral, & has confirmed my conclusions exactly. You sneer too much at our mineralogists here. I expect you are *touched* at being taught a lesson by them.[77]

Eaton was still not convinced.

> If you should see tons of Sul. barytes of S. Hampton, Hatfield, Littlefalls &c. as I have; and then see a little world of the Schoharie mineral; you would say, one would be as excusable for pronouncing quartz and chalk, the same mineral, as he who would pronounce these the same.[78]

The dispute was eventually resolved in a way that satisfied both parties. An abstract of the proceedings of the Lyceum of Natural History published in Silliman's *American Journal of Science* in 1820 contained an account of Torrey's analysis of the mineral, concluding that it should be designated *fibrous sulphate of barytes* but adding: "the fibrous sulphate of baryte analyzed by Klaproth . . . appears to be a very different variety from the mineral in question, and as every new variety should receive a name, we may retain that given to it by Mr. Eaton." [79] Modern analysis confirms that the substance was an impure barite.

---

[76] Eaton to Torrey, July 12, 1819, quoted in McAllister, *Amos Eaton*, p. 267. See also *Amer. Monthly Mag. and Crit. Rev.* 4 (1818–1819) : p. 132.

[77] Torrey to Eaton, July 13, 1819, *ibid.*, p. 267.

[78] Eaton to Torrey, August 14, 1819, *ibid.*, p. 268–269.

[79] "Domestic Intelligence," *Amer. Jour. Science* 2 (1820) : p. 368. For MacNeven's analysis of "Schoharieite" see William James MacNeven, *Chymical Exercises, in the Laboratory of the College of Physicians and Surgeons by the Pupils of the*

Strontianite, barite and calcite all occur together at Schoharie, often intermixed [writes Clifford Frondel]. The troubles of Eaton *et al.* probably arose in that they were dealing with variable mixtures, and probably handed around and collected samples not quite identical. Barite in this type of occurrence also contains more or less strontium in solid solution, adding perhaps to the confusion. The mineral undoubtedly was chiefly barite.[80]

Soon after the controversy over "Schoharieite," Eaton chided Torrey for his conduct in respect to a mineral found by Eaton's colleague and former pupil Ebenezer Emmons in an abandoned iron mine in Richmond, Massachusetts. On Eaton's advice, Emmons sent specimens to Torrey and to Professor Chester Dewey of Williams College for analysis. Dewey concluded that the mineral was a variety of wavellite and described it as such in the *American Journal of Science*, but Torrey was convinced from his own analysis and from the work of the Swedish chemist Berzelius on European varieties of wavellite that the Richmond mineral was a different substance both in composition and in external character.[81] Learning Torrey's views, Eaton suggested that the mineral be called *gibbsite* in honor of Colonel George Gibbs:

Gibbs will always be remembered in this country as the very father of *correct* American Mineralogy. There were some pretty good mineralogists before he began; but his cabinet first set us all off to work, hunting up our own minerals. Besides his own exertions, you know, are unequalled.[82]

Torrey accepted Eaton's suggestion gladly, but when he published his account of gibbsite in 1822 he assigned credit for the discovery of the new mineral to himself and Chester Dewey, mentioning Emmons only as having sent him a "promiscuous collection" of minerals in which the new mineral was subsequently discovered. Eaton was outraged.

Now I assure you [he wrote to Torrey], Emmons is, in all respects, *your superior* as a mineralogist; excepting that he has not given much attention to the chemical analysis of minerals. He is familiar with every mineral in New England, and is considered as a very accurate observer. He first showed the Richmond mineral to me. He told me, at the same time, that it appeared to be new. He gave conclusive and scientific reasons for his opinion. I advised him to send it to you and Dewey. Dewey treated him fairly, and acknowledged him as the discoverer; not a mere fool, who happened to rake up a mass of pebbles . . . as you did. Emmons is the *real and only* discoverer, according to all the rules of the subject. . . .

The truth is, if the countryman has the eye of a Linneus [*sic*] and the science of a Vanquelin [*sic*], every Cockney of a city, where salt water washes the sewers, sets himself up as his superior. Yes! he even persuades himself that a countryman feels himself honoured if he is permitted to make Lord Cockney a present.[83]

Despite their differences, Torrey and Eaton remained friends. The New York scientists joined hands with their colleagues in Albany and Troy in supporting the Lyceum of Natural History and in promoting the establishment of the New York Geological and Natural History Survey. The story of the founding and notable achievements of that survey lies beyond the scope of this book, but it should be noted that the mineralogists and geologists of the Survey recognized their debt to the men who had trained and inspired them. In the "Preface" to the volume on the mineralogy of New York, Lewis C. Beck, brother of Theodric Romeyn Beck and, like him, a graduate of

---

Laboratory, under the Direction of the Professor of Chymistry and Materia Medica, William James MacNeven (New York, 1819), pp. 16–26. MacNeven writes (p. 16): "The mineralogical intelligence of an immense bed of sulphate of strontian having been recently discovered in this state, excited a great deal of interest among the learned of America, and has been announced in Europe; for we find that in the 13th volume of Thomson's Annals, this news is communicated to the numerous readers of that excellent scientific journal. And as an abundant supply of strontites would make a most desirable addition to our mineralogy, I was solicitous to examine the substance thus called, and no less disappointed to find it prove nothing more than a variety of sulphate of barytes. It was made the subject of analytical instruction to my private pupils, two of whom analyzed it in the laboratory of the College." In 1820 Amos Eaton accepted the name *fibrous sulphate of barytes* for the disputed mineral, adding in a footnote: "It has been supposed by some, that this variety is not new. But it has been examined by some of the most accurate European mineralogists, who pronounce it absolutely so." See Amos Eaton and Theodric R. Beck, *A Geological Survey of the County of Albany, Taken Under the Direction of the Agricultural Society of the County* (Albany, 1820), p. 37 n. The authors are grateful to the Beinecke Library of Yale University for providing copies of these works by MacNeven and Eaton and Beck.

[80] Communication from Professor Frondel, Harvard University, to the authors.

[81] John Torrey, "Description and Analysis of Gibbsite, a New Mineral," *New-York Medical and Physical Jour.* 1 (1822): pp. 72–73; Chester Dewey, "Geological Section from Taconick Range, In Williamstown, to the City of Troy, on the Hudson, by Professor Dewey. Williamstown, July 4, 1820," *Amer. Jour. Science* 2 (1820): pp. 246–249. See also *ibid.* 3 (1821): p. 239.

[82] Eaton to Torrey, February 27, 1820, quoted in McAllister, *Amos Eaton*, p. 264.

[83] Eaton to Torrey, August 4, 1822, quoted in *ibid.*, pp. 271–272. Twenty-seven years later Benjamin Silliman, Jr., noted that Torrey's analysis of gibbsite as a terhydrate of alumina had been confirmed by the analyses of Chester Dewey and of Thomas Thomson of Glasgow University, but that Rudolph Hermann of St. Petersburg had shown the similarity between gibbsite and the hydrargillite described by Gustav Rose. (Benjamin Silliman, Jr., "On Gibbsite and Allophane, from Richmond Massachusetts," *Amer. Jour. Science* 57 (1849): pp. 411–417.) Professor Clifford Frondel, a modern authority, writes: "Gibbsite, $Al(OH)_3$, is a quite valid mineral. The original find at Richmond, Berkshire Co., Mass. was as fibrous crusts and stalactites. The mineral is identical with the hydrargillite of Rose 1839 from Slatoust; gibbsite has priority. Rose's material was crystallized and supposedly different from gibbsite, but later shown identical." Wavellite is a hydrous basic aluminum phosphate: $Al_3(OH)_3(PO_4)_2 \cdot 5H_2O$.

the College of Physicians and Surgeons, paid tribute to the pioneering work of Mitchill, Bruce, Akerly, Griscom, the Society for the Promotion of Useful Arts, and the Lyceum of Natural History. In his report on the geology of the Fourth District of New York, James Hall called Mitchill "the father of natural history in New York" and acknowledged his own debt to Amos Eaton, with whom he had studied at the Rensselaer School.

The published record of the researches of the pioneering generation of mineralogists in New York is scattered through a variety of journals, beginning with the *Medical Repository*, Bruce's *American Mineralogical Journal*, and the *Transactions of the Society for the Promotion of Useful Arts*. After 1818 Silliman's *American Journal of Science* provided the main avenue for publication, but mineralogical and geological articles also appeared in such short-lived publications as Biglow's *American Monthly Magazine and Critical Review* (1817–1819), *The Ploughboy* (Albany, 1819–1822), and *The New York Medical and Physical Journal* (1822–1830). These journals and surviving correspondence confirm the picture of enthusiastic activity delineated in Mitchill's letter to Reuben Haines of Philadelphia in April, 1817: "Our young men have formed a Lyceum of Natural History; and hold weekly meetings. They are influenced by a zeal unknown among us until now." Besides Mitchill himself the Lyceum's mineralogical roster included Frederick C. Schaeffer, John Torrey, Peter S. Townsend, Nathaniel Paulding, George Gibbs, James Pierce, John B. Bogert, Joseph Delafield, William MacNeven, and Samuel Akerly among the New York membership. Farther afield were Lewis C. Beck, Theodric Romeyn Beck, and Jeremiah Van Rensselaer.

Collecting minerals was the principal interest of most of these men, but a considerable number of them, including Mitchill, Torrey, Gibbs, Renwick, Paulding, and MacNeven, were also interested in chemical and crystallographic analysis. William MacNeven was an Irish refugee who had received his medical education in Vienna, served in the Irish Brigade of the French army after being arrested for his part in the Irish rebellion of 1798, and settled in New York in 1803, where he became professor of chemistry in the College of Physicians and Surgeons in 1808. His analysis of schoharieite was published in German in Schweigger's *Journal für Chemie und Physik* in 1820, along with his "Analysis of Crystallized Dolomites from North America," selected by the editor as a paper "which distinguishes itself through exactness and skilful development of the result." MacNeven's *Chymical Exercises, in the Laboratory, under the Direction of the Professor of Chymistry and Materia Medica, William James MacNeven*, published in 1819, shows that chemical analysis of minerals formed a regular part of the instruction of his pupils. "The present sheet," he

wrote, "affords a specimen of the mode of proceeding and of what may be expected from young Physicians so instructed, if they should choose, hereafter, to continue their attention to the same pursuits." Among the specimens analyzed by his pupils, two of whom were from South Carolina, were dolomite from the Kingsbridge quarry on Manhattan Island and the schoharieite from Carlisle, N.Y. The geology of the quarry where the dolomite specimens were found was described at some length on the basis of observations by MacNeven's colleague in the Lyceum of Natural History, James Pierce.

The limestone district adjacent to Kingsbridge is shown, by the character of its minerals and the position of its strata, to be chiefly primitive [MacNeven wrote]. The marble extends two or three miles into the county of New-York, and is the termination of a range of primitive granular limestone that commences in Canada, runs through the western parts of New England, and arrives in our neighbourhood through West Chester county. The direction of this range is nearly parallel with the primitive strata of New England and New York, running from northeast to south-west.

The granular limestone of Kingsbridge is stratified, presenting an inclination to the south-east, of from sixty to seventy degrees, as determined by Mr. Pierce. It is crossed by vertical veins of granite and quartz [pegmatite], from a few inches to a foot in width. They extend from the summit of the rock to the lowest depths yet penetrated, dividing the quarries into distinct sections. Mr. Pierce looks upon this circumstance as one that affords additional evidence of the primitive character of this region. Veins of granite having seldom, if ever, been seen elsewhere in limestone, must here be considered as of more recent formation than the rock they traverse. . . .

The limestone of Kingsbridge embraces pyroxene, tremolite, mica, fetid quartz, oxide of titanium, adularia, tourmaline, and sulphuret of iron. The pyroxene, or white augite is mostly in four-sided elongated tables, the sides of various proportions, and nearly rectangular. Its crystals are sometimes found in eight-sided prisms. Kingsbridge was, for a long time, the only locality for this mineral in the United States, but Mr. Pierce has lately met with it at Singsing, higher up the Hudson.[84]

James Pierce also sent specimens to John Torrey, on whom Pierce seems to have relied heavily for the chemical analysis of minerals. Not much is known about Pierce. Listed as "Shipmaster, 42 Water Street" in the Lyceum's list of resident members, he resided for a time in Catskill, N.Y., and helped to organize the Catskill Lyceum of Natural History, of which he became president. Whatever his regular occupation, Pierce seems to have had plenty of leisure for mineralogical and geological research. According to Samuel Akerly, Pierce's "pedestrian excursions" over the country lying between Navesink Hill, New

---

[84] MacNeven, *Chymical Exercises* pp. 1–2, 3. See Desmond Reilly, "An Irish-American Chemist, William James MacNeven, 1763–1841," *Chymia* 2 (1949): pp. 17–26.

Jersey, and Albany were more frequent and extensive than those of any other individual.[85]

In 1819 Pierce read a memoir before the Lyceum of Natural History on the mineralogy and geology of the "secondary region" of New Jersey and New York. Later published in the *American Journal of Science*, the memoir reveals the extent of his field work and of his reliance on John Torrey for the analysis and identification of minerals. After describing the location and contours of the region, Pierce gave an extended account of the mineral content of the formations there. Among the minerals Torrey identified for Pierce were prehnite, kaolin, chlorite, datolite, and an ore consisting of pyritous and green carbonate of copper. The datolite was at first taken for a specimen of prehnite, but Torrey's analysis revealed its true character. Pierce's account is notable:

Before the autumn of 1818, prehnite, calcareous spar, and carbonate of copper [malachite], were the only minerals observed imbedded in the greenstone ranges adjacent to Patterson [sic]. At that period I met with, near the falls, superior specimens of zeolite, stilbite, analcime, and datholite, together with fine masses of prehnite; amethyst has been since discovered by Judge [Charles] Kinsey, and Mr. J. I. Foote, residents of Patterson. To these gentlemen I am indebted for many useful facts. . . . Since my discovery of datholite at Patterson, I have sought in vain for this mineral elsewhere in the greenstone ranges; the vicinity of the falls is the only locality for it yet found in this country and there is but one in Europe; its character was conjectured by Col. Gibbs and ascertained by Dr. Torry [sic] by analysis. The Patterson datholite will probably be regarded as a new variety of the Norwegian mineral, differing in crystaline form and proportion of its constituent parts.[86]

---

[85] Akerly, *Essay on the Geology of the Hudson River*. See also "Extract of a Letter to the Editor from James Pierce, Esq. dated Litchfield, August 22, 1820," *Amer. Jour. Science* 3 (1821) : pp. 236–237. Pierce wrote: "The attention of the well informed residents of Catskill has been of late excited to the study of mineralogy, botany, chemistry and agriculture, and they have recently organized a scientific institution under the name of the Catskill Lyceum of Natural History, composed of between twenty and thirty residents, and as many corresponding members, who evince much zeal, and have formed a small cabinet of minerals and plants. . . . I have occasionally read to this society papers on natural history, dwelling more particularly on mineralogy, exemplifying the remarks by specimens to render the external characters of minerals more familiar. I have endeavoured to induce research by drawing their attention to valuable minerals, which, from the geological character of the river valley, and adjacent districts, may be met with; as silver, lead, copper, plumbago, iron, gypsum, alum, salt, coal, marble and marl. I have found in different parts of the Catskill mountains, extensive ledges richly impregnated with alum, and salt licks in the same region. . . . I have found new localities in the valley of the Hudson for galena, plumbago, and iron. If I make Catskill my residence the ensuing summer, I will explore the Catskill, Shuongunck [sic] and Highland ranges." For other items relating to Pierce, see *Amer. Jour. Science* 1 (1818) : pp. 54–55; 2 (1820) : pp. 366–372; 5 (1822) : pp. 26–33.

[86] James Pierce, "Account of the Geology, Mineralogy, Scenery, &c. of the Secondary Region of New-York and New

In the serpentine rocks of Hoboken, New Jersey, Pierce found horizontal veins of native carbonate of magnesia [hydromagnesite]. When first dug out of the rock with a spoon, Pierce reported, "it was soft, white, and very slightly adhesive, from a little moisture; but, when dry, fell to powder without friction."

The nature of the mineral I conjectured as soon as seen, and treated it with diluted sulphuric acid, in which it entirely dissolved with effervescence, forming a bitter fluid and leaving no sediment. Upon evaporation, well-defined crystals of Epsom salts were formed. It differs little from the manufactured carbonate of magnesia of the shops; but is rather a super than a sub-carbonate. It has been analyzed by Professor Mitchill, who found it exclusively composed of magnesia and carbonic acid. . . . When I first mentioned the discovery to mineralogists they were incredulous, supposing it did not natively exist in this state, but I convinced them by uniting it with sulphuric acid.[87]

Commenting on Pierce's discovery in the *American Journal of Science*, Benjamin Silliman noted that it had been made in the same rocks where Archibald Bruce had found magnesium hydroxide. "Mr. Pierce's discovery is not less interesting," Silliman added, "and we presume he will be deemed correct in the opinion, that pure native carbonate of magnesia has not been discovered before. The serpentine of Hoboken, then, is memorable for affording these two new species."[88]

Torrey was also involved in the controversies surrounding a mineral which Colonel Gibbs named *brucite* in honor of his friend Archibald Bruce. According to Torrey, specimens of the mineral in question were first noticed by Bruce in 1811 in a crystallized carbonate of lime collected in Sussex County, New Jersey. Assuming that the substance was an ore of titanium, Bruce sent specimens to the Abbé Haüy, who placed them in his mineral cabinet without undertaking an analysis. Bruce's pupil and laboratory assistant William Langstaff made repeated analyses, however, and found that the substance contained silica, magnesia, and fluoric acid. Learning of Langstaff's experiments, Colonel Gibbs concluded that the mineral was a new one and proposed, with Langstaff's consent, to name it *brucite*. A considerable quantity of the mineral was then sent to Europe under that name.

Meanwhile the mineral had also become known as fluate of magnesia, and Professor Parker Cleaveland listed it as such in his *Elementary Treatise on Mineralogy* (1816), although the account he had hoped to publish of it did not reach him in time to be in-

---

Jersey, and the Adjacent Regions," *Amer. Jour. Science* 2 (1820) : pp. 191–192.

[87] "Carbonate of Magnesia, and Very Uncommon Amianthus, Discovered near New-York—Extract of a Letter from Mr. James Pierce to the Editor, New York, May, 1818," *Amer. Jour. Science* 1 (1818) : pp. 54–55. "Specimens from the Hoboken locality (long since obliterated) are found in many old American collections." (Frondel)

[88] *Ibid.*, p. 55.

cluded. In 1819 Torrey analyzed the mineral, probably at Cleaveland's urging, and reached the tentative conclusion that it was a fluosilicate. But before he could confirm his analysis, he learned of Langstaff's earlier experiments and decided not to publish his own, although he sent Cleaveland a brief description of the mineral under the name brucite for Cleaveland's second edition (1822). Meanwhile the Swedish chemist Berzelius had come across the specimens in Haüy's cabinet and had concluded from his analysis of them that the mineral was identical with one described by Count D'Ohsson in 1817 under the name chondrodite. Since neither D'Ohsson nor Berzelius found any fluoric acid, however, Colonel Gibbs and others in America were inclined to insist that brucite was different from chondrodite.[89]

In Philadelphia yet another bitter controversy was being waged over the same substance. Nuttall's description of the yenite and chondrodite he had found in Sussex County had been criticized by his colleagues in the Academy of Natural Sciences, and Nuttall had left Philadelphia for New York, where he published his paper in the *New York Medical and Physical Journal,* doubtless with the counsel and assistance of his friend John Torrey. Henry Seybert had analyzed Nuttall's "chondrodite" and, finding that it contained fluoric acid, declared it a new mineral species, which he named *maclurite* in honor of William Maclure. An unpleasant controversy between Seybert and Nuttall ensued in the pages of the *American Journal of Science,* Seybert accusing Nuttall of claiming credit for discovering fluoric acid in the substance, Nuttall retorting that the New York mineralogists had known this fact for more than a decade. Nuttall also called attention to the similarity between the mineral and another mineral from Vesuvius he had seen in the cabinet of William Wagner, Jr., in Philadelphia—a mineral known in Europe as humite. To his delight Nuttall found a similar specimen from Vesuvius in the Gibbs collection at Yale.

We have tried it and found it infusible, and are going on further with the comparison [he wrote to Torrey]. It appears that Master Seybert, without consulting his superiors (i.e., ourselves) had wickedly and maliciously christened the Brucite, maclurite, but we are ahead of him.[90]

To resolve the controversy Torrey sent specimens

of the chondrodite to Professor Thomas Thomson at Glasgow University, asking him to analyze them. Thomson's analysis yielded results very similar to those obtained by Seybert, but since it was known by this time that the chondrodite of Count D'Ohsson and Berzelius contained fluoric acid, it now seemed clear that there was but one mineral variously called chondrodite, brucite, and maclurite. Since Count D'Ohsson had been the first to describe and name the substance, Thomson adopted his name, chondrodite. In his "Notes" to Thomson's article, however, Torrey pointed out that chondrodite had recently been shown to be identical in nature with the humite of the Count de Bournon, described in the same year (1817) that D'Ohsson described chondrodite.[91] Bruce's claim to the distinction of having a mineral named in his honor was satisfied in 1827, when François S. Beudant proposed the name brucite for the magnesium hydroxide Bruce had described in 1810. The mineral still bears his name.[92]

Chrondrodite was but one of many minerals sent to Thomas Thomson by Torrey and Nuttall. Thomson's report on these minerals in the *Annals of the Lyceum of Natural History* in 1828 is an interesting example of relations between European and American science at this time. Nuttall was primarily a collector and, as such, stood in the traditional relationship of purveyor of the raw materials of science to the scientists of Europe. Describing his role and the energy and acuity with which he performed it, his biographer has written:

He arduously brought back much new material from remote places for the study of specialists. His keenness as a mineralogist becomes more apparent when the number of his acceptable new varieties is compared with those detected by American contemporaries who were exclusively mineralogists and chemists. The only worker of his active period in this field who exceeded him in the quantity of new forms published was his protégé Charles U. Shepard.[93]

Among the minerals Nuttall sent to Thomson were

[89] Thomas Thomson, "Chemical Examination of Some Minerals, Chiefly from America. By Thomas Thomson, M. D., F. R. S. . . . With Notes by John Torrey, M. D. Professor of Chemistry and Botany in the University of the State of New York," *Annals Lyceum of Natural History of New York* 2 (1828) : pp. 56–61.

[90] Thomas Nuttall to John Torrey, July 2, 1822 (Historical Society of Pennsylvania), quoted in Jeannette E. Graustein, *Thomas Nuttall Naturalist* (Cambridge, Mass., 1967), p. 169. On pp. 163–171 Graustein gives a full account of Nuttall's mineralogical activities in this period, including his controversy with Seybert and his relations with Torrey.

[91] Thomson, "Chemical Examination of Some Minerals . . . With Notes by John Torrey," p. 60. According to Professor Clifford Frondel, Harvard University: "The Chondrodite Group of minerals includes norbergite, chondrodite, humite and clinohumite. They all closely resemble each other, with closely related compositions, and can not be distinguished by qualitative chemical tests. Both norbergite and chondrodite occur in the Franklin marble formation of Sussex County and Orange Co., N.Y.; most of it is norbergite, but this was not recognized until 1928, the minerals earlier being confused as chondrodite. Norbergite was not discovered, as a new mineral, until 1926, in Sweden. Thomson's and Seybert's descriptions are not good enough to tell whether they had norbergite or chondrodite."

[92] François S. Beudant, *Traité Élémentaire de Minéralogie* (Paris, 1824), pp. 487, 838. Beudant describes "hydrate de magnésie" on p. 487. In the index, p. 383, he lists it as "Brucite, ou hydrate de magnésie."

[93] Graustein, *Thomas Nuttall Naturalist,* p. 166.

specimens of diphosphate of iron, manganesian iron ore, bucholzite [fibrolite], xanthite [variety of idocrase], phyllite, pipestone [muscovite and pyrophyllite], ceylanite [variety of spinel], and nuttallite [variety of scapolite], the last of these named in Nuttall's honor by the English mineralogist Henry J. Brooke.

Nuttall judged the nature of his specimens by their external characters, but his friend Torrey aspired to proficiency in the chemical and crystallographical analysis of minerals. From his own and Thomas Thomson's comments it appears that Torrey's views on the nature of the specimens he sent to Thomson were frequently different from Thomson's. Thus, Torrey's rhomboidal silicate of zinc became in Thomson's analysis silicate of manganese, Torrey's manganesian feldspar became ferro-silicate of manganese [impure rhodonite], Torrey's "modification of franklinite" became sesquisilicate of manganese, and Torrey's "new variety of garnet" became idocrase. On the other hand, Torrey's brown manganesian garnet [manganoan andralite] was confirmed by Thomson, and his gibbsite, though not analyzed by Thomson, was confirmed by other analysts and continues to bear that name. It should be added in Torrey's defense that Thomson's judgments were not infallible. Charles Palache, a modern mineralogist, characterizes Thomson's analyses of minerals from the Sparta region of New Jersey as "mostly poor," pointing out that many of them were revised and corrected by Rudolph Hermann in 1849.[94] Mineral analysis was a difficult branch of chemistry, as Edgar Fahs Smith observed in 1926:

. . . chemists have not succeeded in determining the constitution of many minerals—true chemical compounds—as has been done, for example, in the case of salicyclic acid, indigo, and many other organic products. The problem is a difficult one. It calls for an almost superabundance of patience, for a thorough knowledge of a great list of elements, for the power of analytic generalization, etc. . . . Comparatively few chemists have had the courage and patience requisite to search for adequate methods of purification of the material offered by mineral chemistry. Some few methods, falling perhaps within the domain of physical chemistry, have been applied, but there it has ended.[95]

Many of the new minerals sent to Thomson by Torrey and Nuttall came from Sussex County in New Jersey. The mineral riches of this region were easily

available to the mineralogists of both New York and Philadelphia, and the competition to collect and describe them was keen. Honors were fairly evenly divided between Bruce, Pierce, Langstaff, and Torrey on the one side and Henry Seybert, Vanuxem, Keating, Troost, and Nuttall on the other, with perhaps some advantage in favor of the Philadelphians. Philadelphia could boast more mineralogists trained abroad and a more secure natural history establishment in the Academy of Natural Sciences than New York. But the Lyceum of Natural History of New York made up in energy and zeal what it lacked in financial resources, and the College of Physicians and Surgeons did more to interest and train students in mineralogy than any academic institution in Philadelphia. Despite the lack of a cultural tradition favorable to science and learning, the New York mineralogists had made a good beginning. Their efforts were soon to yield substantial results in the Geological and Natural History Survey of New York (1836) and its major contributions to the development of the earth sciences.[96]

V. FIRST EFFORTS IN THE CAMBRIDGE-BOSTON AREA

The development of the sciences in New England was more closely associated with the liberal arts colleges and less dependent on the growth of urban centers than in Pennsylvania and New York. Astronomy flourished at Harvard College during the long tenure (1738–1779) of John Winthrop IV as Hollis Professor of Mathematics and Natural Philosophy, but Boston was slow to emulate Philadelphia in establishing institutions for the promotion of science. The American Academy of Arts and Sciences was founded as a counterpart of the American Philosophical Society in 1780, but Daniel Bowen's museum of curiosities was no match for Peale's museum, and there was no chemical society in Boston to rival the Chemical Society

[94] Charles Palache, *The Minerals of Franklin and Sterling Hill, Sussex County, New Jersey,* U. S. G. S. P. P. 180 (Washington, D.C., 1935), p. 3.

[95] Edgar F. Smith, "Mineral Chemistry," pp. 69–88 in: Charles A. Browne, ed., *A Half Century of Chemistry in America 1876–1926. An Historical Review Commemorating the Fiftieth Anniversary of the American Chemical Society,* separately paged (pp. v-254) in *Jour. Amer. Chem. Soc.* 48 (1926), insert following p. 2014. The quotation is from p. 69. Professor Frondel remarks: "Smith's remarks are somewhat overdrawn, and the situation today is indeed very different."

[96] A summary view of the minerals discovered in the state of New York to 1824 may be found in H. H. Webster, *A Catalogue of the Minerals, Which Have Been Discovered in the State of New-York; Arranged under the Heads of the Respective Counties and Towns in Which They are Found* (Albany, Printed by Websters and Skinners, 1824). The source of the description of each mineral is given. These sources include Bruce's *Journal,* Silliman's *Journal,* the second edition of Cleaveland's *Elementary Treatise,* the geological surveys of Albany and Rensselaer counties and of the vicinity of the Erie Canal by Amos Eaton and his associates, Dr. John H. Steel's report on the geology and mineralogy of Saratoga county in his *Analysis of the Mineral Waters of Saratoga and Ballston . . . with a Geological Map* (2nd ed., Albany, D. Steele, 1819), *A Catalogue of the Minerals, Contained in the Cabinet of the Late Benjamin De Witt, Consisting of More Than Eleven Thousand Specimens* (Albany, G. J. Loomis & Co., 1820), and a manuscript "Catalogue of the Minerals Contained in the Cabinet of the Albany Institute." Webster was a member of the Albany Institute, which was formed by the union of the Society for the Promotion of Useful Arts and the Albany Lyceum of Natural History.

of Philadelphia. Natural history did not achieve a solid institutional base until the founding of the Boston Society of Natural History in 1830, eighteen years after the establishment of the Academy of Natural Sciences of Philadelphia and thirteen years after the Lyceum of Natural History began operations in New York.

A medical school was added to Harvard College in 1782, but it did not amount to much until after its move from Cambridge to Boston in 1810. Although the College itself was the leading liberal arts college in the nation, it was overshadowed by Yale in chemistry, mineralogy, and geology after Benjamin Silliman began the spectacular development of these sciences there in 1806. Even Bowdoin College, a small new college in Maine, surpassed Harvard in the study of earth sciences as a result of the heroic efforts of Parker Cleaveland, who developed his interest in mineralogy after graduating from Harvard. Despite the efforts of a few enthusiasts, mineralogy made slow headway in the Cambridge-Boston area.

As at the University of Pennsylvania, mineralogy entered the curriculum at Harvard by way of the School of Medicine. In 1782 Benjamin Waterhouse, recently returned from his medical studies in Edinburgh, London, and Leyden, was appointed professor of the theory and practice of physic and soon afterward was authorized to deliver a course of lectures on natural history in addition to his medical lectures. This course included material on mineralogy from its inception, but Waterhouse's interest in minerals was greatly stimulated by a generous gift of specimens from his British friend Dr. John Coakley Lettsom, a liberal patron of American science. In 1793 Lettsom sent Harvard College a collection of several hundred minerals, subsequently augmented by further gifts. Describing some of these acquisitions, Waterhouse wrote:

A hundred fine specimens from the Spanish mines have just arrived from the same gentleman. Here are some specimens of *gold ore*, a great variety of *silver*, a still greater of *copper, iron, tin, lead, zinc, antimony, arsenic, bismuth, cobalt, nickel, manganese*; not to mention innumerable *spars, fluors, chrystallizations* [sic] *petrifactions, salts,* and *saline earths:* with mixtures and combinations of each, forming a very useful and splendid collection. These Minerals were collected from Mexico, different parts of Germany, from Transylvania, Hungary, and Poland, as well as from Turkey, Russia, Sweden, Denmark, Norway, Great-Britain, Ireland, Italy, and elsewhere.[1]

In 1795, while arrangements were being made to display these treasures in a suitable cabinet, 189 additional specimens were given to the College by the Republic of France "as samples of the riches of the French soil," with a request that "you Will acquaint us of the Situation of the part of the Soil of the United States which you have had an opportunity to examine, and . . . entertain an Exact correspondance [sic] with us which will without doubt be of an Advantage for our Instruction." "Your example," the directors of the Agence des Mines added, "has Influenced our Souls with the Sacred fire of Liberty; as you, We have felt that Arts was necessary to maintain it, and Metals to deffend it; as Us you will think that the reunion of sentiments, The propagation of Knowledge, the perpetual Exchange of Instructions are to assure the hapiness of the world." [2]

It seems unlikely that the libertarian sentiments of French revolutionaries inspired much enthusiasm among Bostonian Federalists, although Waterhouse himself, the lone Jeffersonian on the medical faculty, doubtless appreciated them. In any case, the donation was accepted with suitable acknowledgments, and the mineral collections were soon on display in "an elegant mahogany Cabinet, eighteen feet long and from ten to twelve high, placed in the Philosophy-Chamber . . . for the inspection of the curious." They constituted, Waterhouse declared proudly, "by far the richest and most extensive collection of minerals in the United States."

Both the English and French collection happened to be more deficient in Italian *marbles,* and *volcanic lavas,* than in almost any other fossil [Waterhouse added], which deficiency has been generously supplied by the Hon. Mr. [James] Bowdoin, who has presented the Cabinet with an *hundred and fifty* specimens of those two productions. . . .
These minerals are arranged (with but very few exceptions) in systematic order: Each article is numbered, which numbers answer to those of a descriptive catalogue, which has been carefully made out for public inspection; for besides the name of the Mineral and the place it came from, the opposite page contains definitions and explanatory notes; an addition not wholly superfluous in a region where the science of Mineralogy is but in its infancy.[3]

E. Graustein, "Natural History at Harvard College, 1788–1842," *Proc. Cambridge Hist. Soc.* 38 (1961) : pp. 69–86.
[2] Quoted in Cohen, *Some Early Tools of American Science,* pp. 118–119. See *Journal des Mines* 1, 5 (January, 1795) : p. 95, and 3, 15 (November, 1795) : pp. 52–54 for notices of the gift of minerals and the reply from President Joseph Willard.
[3] Waterhouse, "Cabinet of Ores and Other Minerals," p. 193. Waterhouse remarks that the London merchant Thomas Hollis, who established the Hollis Professorship of Mathematics and Natural Philosophy at Harvard, wrote in the fly leaf of a book on mineralogy which he sent to the College in 1768: "A Professorship of Chemistry and Mineralogy, to be instituted at Harvard College, which alone would, it is apprehended, bestow wealth on New-England, *with maintenance of its industry,* cannot be too much recommended to the gentle-

[1] Benjamin Waterhouse, "Cabinet of Ores and Other Minerals, in the University of Cambridge, in New-England," quoted in full in Thomas J. Pettigrew, *Memoirs of the Life and Writings of the Late John Coakley Lettsom . . . With a Selection from His Correspondence* (3 v., London, 1817) 1: p. 192. See also I. Bernard Cohen, *Some Early Tools of American Science: An Account of the Early Scientific Instruments and Mineralogical and Biological Collections in Harvard University* (Cambridge, Mass., 1950), pp. 120–121; Jeannette

To aid the study of these specimens, the College acquired an assaying apparatus in 1797, complete with goniometer, blowpipe, glass retorts, pestle and mortar, etc., but although he continued as Keeper of the Mineral Cabinet until 1810, Waterhouse did not make a significant contribution to the development of mineralogy in America apart from his general lectures on natural history. Letters to Dr. John Lettsom and the outline of the natural history lectures in the pages of *The Monthly Anthology and Boston Review* provide glimpses of his views on mineralogy. In a letter to Lettsom dated October 18, 1799, he wrote:

Your donation of minerals has restored the title of Mineralogist to its honourable rank, so that some of our young gentlemen have chosen that branch of natural history as a theme for their public declamations. I was less acquainted with that department of natural history than perhaps any other. I was, however, determined to master the subject: I began with Dr. Hill, and then read those books you sent me, namely, Cronstedt, Kirwan, and Schmeisser. I was underground the greatest part of the winter and did not emerge into light till about the time the frost came out of the ground, and then I composed a few lectures, and distributed a printed letter, describing our collection of minerals, and inviting men of science and others, to send everything above the appearance of a common stone to our mineralogical museum.

Its consequences exceeded my most sanguine expectations. Some farmers quitted their agriculture and came above 100 miles to bring me what they conceived to be gold ore, which was only splendid pyrites. I undeceived these people in as gentle a manner as I could, by telling them that the very bowels of the earth gave man lessons of wisdom. . . . Then I showed them true gold and silver ores from your collection, and contrasted them with the worthless but glittering pyrites. . . .[4]

Among the "Heads of Lectures" (1800–1804) published by Waterhouse in *The Monthly Anthology* only two heads pertain directly to mineralogy and geology:

XVII. MINERALOGY. The contents of the earth but little known:—all below *three thousand* feet is dark conjecture. *Mountains* distinguished into *primaeval* and *alluvial*. The first are the "everlasting hills," which never contain metallic ores, nor petrefactions [*sic*], nor any animal exuviae; of this kind are, the *Alps* and *Pyrenees*, in Europe; . . . and the *Andes*, in America. These *preceded* the formation of vegetables and animals. The second are as evidently of *posterior* formation. They lie in *strata*, contain ores, petrefactions of vegetables, and vestiges of organic substances. These *alluvial* mountains formed at, or since, the deluge; the primaeval as old as the globe. *Kirwan* recommended.

XVIII. THE MINERALOGICAL SCHOOLS of Sweden, Germany, and France. . . . *Cronstedt* recommended.

The LETTSOMIAN *Cabinet of Minerals.* Mineralogy of more importance, at the present, to AMERICA, than

Botany. We are dependent on foreign nations for riches that lie under our feet! . . . The extravagant price set on diamonds, and other glittering stones, ridiculous in the eyes of REPUBLICANS. . . .[5]

This outline suggests that Waterhouse's lectures on mineralogy and geology were distinguished more for patriotism and utilitarianism than for mastery of the latest developments in earth science. "I have been the *pioneer* in this business," he wrote Lettsom, "and having broken the way, must leave to my successor the easy task of smoothing it." When young Benjamin Silliman visited Boston and Cambridge in 1807 in company with Colonel George Gibbs of Newport, Rhode Island, he admired the "small but beautiful" collection of minerals at Harvard College but concluded that mineralogy had not really taken root there. "There was not at this period . . . much of a spirit of science in Boston," he recalled in his "Reminiscences." [6]

In the summer of 1807, however, the attention of the Boston literati was drawn to mineralogy by the arrival from Paris of Silvain Godon, whose mineralogical career in Philadelphia has already been described. Godon was warmly received in Boston and urged by the Reverend Joseph Buckminster and others to undertake a course of public lectures on mineralogy. Godon agreed and, after a trip to Philadelphia, settled his "petite famille" in Boston and began to plan his lectures.

I have left the countryside in Brookline to establish myself in Boston [he wrote to John Vaughan in November]. I have taken a house which I am now busy arranging to receive my auditors. The lectures which I am preparing will put me in a position to appreciate the genuine taste which the Bostonians have for useful sciences. I do not know yet what the number of my auditors will be; they have given me hope that the number will exceed 50.[7]

As things turned out, the Reverend Buckminster's estimate of Godon's probable audience was much too high. Although Godon announced that he would admit impecunious young students to the lectures free of charge, the response was disappointing.

My offers were without success [he wrote to John Vaughan]. I often saw persons unknown to me present at my lectures, but it was only for a lecture or two at the most. A revolution in attitudes such as that imagined by by M. B. [Buckminster] is not likely. I am far from thinking that it will always be thus, however, and I do not see why the country which gave birth to Dr. Franklin would be one to disdain *la bonne philosophie*.[8]

---

men there, as individuals and legislators" (p. 195). The Bowdoin gift of marbles (polished slabs mostly three or four inches across) is still preserved at Harvard.

[4] Pettigrew, *Memoirs of . . . John Coakley Lettsom* 2: pp. 466–467. See also James J. Abraham, *Lettsom, His Life, Times, Friends and Descendants* (London, 1933), pp. 360–361.

[5] "Heads of a Course of Lectures Intended as an Introduction to Natural History. By B. Waterhouse, M. D. . . ," *Monthly Anthology and Boston Rev.* 3 (1806): pp. 311–312.

[6] Quoted in Fisher, *Life of Benjamin Silliman* 1: p. 19.

[7] Silvain Godon to John Vaughan, Boston, November 30, 1807, Library of the American Philosophical Society. Original in French.

[8] Godon to John Vaughan, Frankfort, Pennsylvania, October 18, 1808, Library of the American Philosophical Society.

According to *The Monthly Anthology*, Godon gave about thirty lectures illustrated with experiments and with specimens from his cabinet of minerals. He was invited to publish the lectures but declined on the ground that they were too elementary. The best way of promoting the study of American mineralogy, he told his well-wishers, would be to establish a public collection of minerals in Boston.[9]

Although his lecture series was no great success, Godon took advantage of his stay in Boston to explore the mineralogy and geology of the region, venturing as far north as Maine, where he was cordially received by Cleaveland at Bowdoin and by Cleaveland's close friend Benjamin Vaughan, who resided at Hallowell. Greatly impressed with the scientific ardor of these two men, Godon maintained correspondence with them for some time after his departure for Philadelphia in 1808. In 1809 the results of his mineralogical explorations in New England were published in the *Memoirs* of the American Academy of Arts and Sciences under the title "Mineralogical Observations Made in the Environs of Boston, in the Years 1807 and 1808," communicated for Godon by Judge John Davis, a leading patron of science in Boston.[10]

Godon noted that few publications or descriptions of mineral localities had been attempted, and nearly all of these restricted to Europe. Only when mineral localities had been described for all parts of the globe, he declared, would it be possible to construct a universal mineralogical map showing at a glance the mineral riches of the entire earth. Godon viewed the mineralogy of the Boston region from the standpoint of descriptive geology. In a tabular view he listed a number of "simple minerals" and "aggregate minerals," most of the former now termed minerals, and the latter rocks. With respect to "aggregate minerals," he defined amphiboloid (diorite and diabase) as an aggregate of amphibol and feldspar, containing quartz, epidote, talc, mica, and sulphurated iron, and explained that when feldspar was in the greatest proportion it was called "felsparoid" [granite]. The buildings of the town of Boston, he wrote, rested on alluvial deposits containing a large proportion of clay,

which protected the town from the encroachment of the ocean and was responsible for the presence of many springs of pure fresh water. West of Boston the rock was "amphiboloid passing frequently into felsparoid," whereas north of Boston felsparoid predominated. Simple petrosilex [felsite] occurred in Quincy and Dorchester, and porphyritic petrosilex in Malden and Lynn. Argilloid rock [argillite] was found in Brighton, Newton, and Roxbury, while wacke [conglomerate], known as plum pudding stone by the inhabitants, spread from the northeast to the southwest. Godon described the Blue Hills in Milton as felsparoid, verging to petrosilex both simple and porphyritic, and noted that the rock was interrupted by large veins of quartz, which sometimes contained cavities. Although he gave a fairly good description of the geological terrain of Boston and vicinity, his mineralogical account was less satisfactory than that provided by the Danas ten years later.

Discussing Godon's attempts at mineralogical analysis, Clifford Frondel has written:

Godon's analysis of the rock now known as the Cambridge argillite is a naive effort, even in the context of its times, and betrays a thorough lack of understanding of both general chemistry and of analytical methods. It doubtless is not representative of what American chemists of the day could do, as indicated by Silliman's analysis of the Weston meteorite, made in the same year. . . . Godon's statement (page 147) that "This analysis . . . is sufficient to establish the important fact of the existence of potash and soda as elements in some rocks in this part of the world" is painfully revealing, especially since he mentions feldspar and feldspathic rocks in the Boston area elsewhere in his paper. . . .
His analytical method consisted of letting a powdered sample of the rock sit in sulfuric acid for 15 days, and then making some very simple operations on the 15 per cent of the sample that dissolved; he then assumed that the 85 per cent residue contained 55 per cent SiO_2, as representative of such rocks, and that the balance had the same composition as that found in the dissolved part![11]

The 1809 appointment of Dr. John Gorham as adjunct professor of chemistry and materia medica at "the University in Cambridge," as Harvard College was called in those days, provided more lasting stimulus to the development of mineralogy in the Boston area than Godon's visit. A graduate of the Harvard School of Medicine, Gorham had recently returned from his medical studies in Paris, London, and Edinburgh. At Edinburgh he shared rooms with Benjamin Silliman and attended lectures on chemistry, mineralogy, and geology as well as medicine. The two young men returned to America on the same ship and corresponded with each other for many years afterward. Silliman introduced chemistry and allied subjects at Yale, and Gorham was soon asked to teach

[9] *Monthly Anthology and Boston Rev.* 5 (April, 1808) : pp. 209–211.

[10] Silvain Godon, "Mineralogical Observations, Made in the Environs of Boston, in the Years 1807 and 1808," *Memoirs Amer. Acad. Arts and Sciences* 3, 1 (1809) : pp. 127–154. The authors are grateful to Professors Marland P. Billings and Clifford Frondel of Harvard University for their commentaries on this article. Dr. John Collins Warren credited Godon with having awakened an interest in mineralogy and geology in him: "But for the mineralogy of Godon, I should not have been able to advance much in this study. . . . He instructed us by lectures, and by labor in the field. Once or twice a week, we mounted our horses, and, scampering over the country, were led by him to such places as he thought interesting for geological investigations." See Edward Warren, *The Life of John Collins Warren, M. D. . . .* (2 v., Boston, 1860) 2 : pp. 9–10.

[11] Letter from Clifford Frondel to John C. Greene, Cambridge, Massachusetts, February 1, 1972, quoted with the kind permission of Professor Frondel.

his newly acquired knowledge of these subjects to the medical students and undergraduates at Harvard as an assistant to the Erving Professor of Chemistry and Mineralogy, Dr. Aaron Dexter, to whose chair he succeeded in 1816. Gorham accepted the appointment with enthusiasm and was soon writing to Parker Cleaveland at Bowdoin College about his plans for the development of chemistry and mineralogy at Harvard.

My Hallowell friends have supplied me with only a few specimens of staurotide [staurolite] [he wrote in October, 1809]. I have wished for an opportunity of corresponding with you and of exchanging specimens as it would doubtless prove of mutual advantage. Some of my friends, particularly [Dr. William] Meade and Mr. Selden have furnished me with some from your neighborhood, but in general they have been such [as I?] would not wish to constitute a part of my cabinet. If you will do me the favor to supply me with others, I shall be happy, in return, to make up a box of foreign materials such as my cabinet will afford. I am also frequently receiving native specimens and I shall send such of them as I think may be useful or interesting to you.[12]

Gorham informed Benjamin Silliman that between sixty and seventy students were attending his chemical and mineralogical lectures at Harvard and that plans were under way to establish a mineralogical society in Boston, displaying as a prize exhibit the mineral collections of Colonel George Gibbs.[13] Colonel Gibbs had moved to Boston from Newport, Rhode Island, about this time and was involved in an ambitious scheme to form a mineralogical society and establish a school of mines, using his collections for this purpose. Parker Cleaveland described some aspects of the project to Benjamin Silliman several years later:

About 2 or 3 years since Col. G. & several other mineralogists in Boston made considerable progress in the formation of a mineralogical society in that town. Public rooms were to be provided, Col. G's collection deposited; and several courses of Public lectures given on Min. Geol. &c. The gentlemen, who were to lecture on the several branches, were even named. The substance of the above was communicated in a letter from Col. G. to myself. Here the business slept, and, for a long time, I heard nothing of it. Sometime after, when enquiring of some Boston gentlemen concerning the welfare of their society, they informed me, that all had fallen through, and appeared inclined to attribute the failure to Col. G. The particulars of this failure, however, were either not related to me, or I have forgotten them.[14]

Despite the collapse of this project and Colonel Gibbs's subsequent decision to place his collection on display at Yale College, Gorham's enthusiasm for mineralogy grew. "My mineralogical cabinet increases daily, I find the study of this science becomes every day more interesting and I recollect with pleasure that I derived the first pleasant impressions of it from you, while we were together at Edinburgh," he wrote to Silliman in September, 1810. "It is really surprising," he told Silliman nine months later, "how very suddenly a taste for mineralogy has sprung up among us. Three years ago there was scarcely a man in Massachusetts and perhaps in the United States who could distinguish one stone from another; now every body here is collecting a cabinet or engaging in mines."[15]

As time went on, however, Gorham found it increasingly difficult to combine the study of mineralogy with medical practice and lecturing on chemistry at the School of Medicine in Boston and at the College in Cambridge. Field work was all but impossible. "My cabinet increases very slowly," he wrote to Silliman, "for I cannot devote sufficient time to this pursuit without encroaching on the duties of my profession, which from principle as well as interest must not be thought of."[16] Meanwhile the Harvard Corporation, doubtless impressed by the rapid progress of mineralogy and geology at Yale under Silliman's leadership, was considering ways of establishing these subjects in the undergraduate curriculum at Harvard. Confidential word of these discussions reached Parker Cleaveland through Levi Hedge, professor of logic and metaphysics, in March, 1815:

It has long been a desideratum to have chemistry and mineralogy taught here as part of our collegiate course, and there has been heretofore considerable talk of a professor of these branches, who should reside at the College. The return of peace I understand has revived the subject, and the Corporation I believe are making arrangments relative to this object, It is doubtful whether either of the present chemical professors would be willing to remove to Cambridge. Dr. Dexter is rather too far advanced to contract new habits; and Dr. Gorham has too great an extent of practice to be willing to give it up. If this be the fact the Corporation will be compelled to look to some other quarter. I know not whether your name has been mentioned, and if it have [sic], they would naturally doubt your willingness to sacrifice the ties, which bind you to your present situation. Will you have

[12] John Gorham to Parker Cleaveland, Boston, October 23, 1809, Cleaveland Correspondence, New York Public Library.

[13] John Gorham to Benjamin Silliman, Boston, November 20, 1809, Silliman Correspondence, Beinecke Library, Yale University.

[14] Parker Cleaveland to Benjamin Silliman, Brunswick, Maine, May 21, 1813, Cleaveland Correspondence, Bowdoin College Library. Some slight further light is thrown on this project by two other letters. On November 6, 1809, Parker Cleaveland wrote to Gibbs [Gibbs Correspondence, Newport Historical Society]: "The following extract of a letter from Majr Tilden of Boston will explain my motives in introducing the subject at this time. 'We are about establishing a min-

eralogical society in this town (Boston) and probably shall have Col. Gibbs collection deposited with it.'" Also in the Gibbs Correspondence is a letter from François Gillet-Laumont, dated Paris, February 25, 1811, which remarks: "Vous désirez établir une École des Mines à Boston, c'est un beau projet et si je puis vous y être utile vous pouvez disposer de moi."

[15] John Gorham to Benjamin Silliman, Boston, September 28, 1810 and June 20, 1811, Silliman Correspondence, Beinecke Library, Yale University.

[16] Gorham to Silliman, Boston, June 20, 1811, Silliman Correspondence, Beinecke Library, Yale University.

the goodness to let me know your feelings on the subject?[17]

In due course Cleaveland was offered a position at Harvard, but he decided not to accept it.[18] Meanwhile Dr. John W. Webster, who was to succeed Gorham as Erving Professor of Chemistry and Mineralogy in 1827, was busy seeking an appointment in mineralogy and geology. Having returned from his studies in Europe with an extensive collection of minerals, he began negotiations for Harvard College to purchase them or to accept them on deposit with Webster as a lecturer on mineralogy and geology. Webster's ambitions and the confused state of the earth sciences at Harvard are reflected in a letter from Webster to Benjamin Silliman in November, 1819:

The business of the cabinet has pretty nearly come to an issue, that is, my proposal made to Cambridge to purchase my minerals, or to allow me to deposit them the College paying me 600 Dls a year with the privilege of lecturing in Cambridge or Boston as a class would be made up, was referred to the President & Judge Davis, who give out that the College is *too poor* at present. From a conversation with Dr. Gorham I find him anxious to have a separate teacher of mineralogy & he is desirous of relinquishing that part of his duty & offers very handsomely to give up part of his salary. The University however I am induced to think from the observations of Dr. Kirkland have strong hope of inducing Col. Gibbs to remove his cabinet from Yale to Cambridge, they say you "have had it long enough," I mention this to you "sub rosa" that you may (if there is a doubt as to its being finally given to Yale) correct any mistaken hopes of our friends at Cambridge pretty publicly; as there is a disposition on the part of the President to have a cabinet, if he has no hopes of persuading the Col. to change the place of deposit of his, I have some hope he will endeavour to purchase mine. Col. G. has you know many connections in the University here & they of course will use all their influence, & perhaps if they find he will not deposit, they will find money to purchase it, in truth the President hinted as much & that there were some arrangements in favor of a former officer of the College now in Europe as to mineralogy.[19]

Apparently there was some substance to Webster's allegations, for Gorham had written Silliman two years earlier that "we are about to commence a collection at Cambridge and endeavour to outnumber the cabinet at Yale." He did not suggest, however, that an attempt

would be made to acquire Colonel Gibbs's cabinet, then on display at Yale.[20]

Webster was unsuccessful in his efforts to persuade Harvard College to buy his collections, and the Gibbs collection remained at Yale. In 1820 Dr. Andrew Richie presented Harvard College with an extensive collection of minerals purchased from C. B. Blöde, a Dresden chemist and mineralogist, and in 1824 several thousand more specimens were bought for the Mineral Cabinet by some wealthy Bostonians. Webster arranged the collections in scientific order and was appointed lecturer on chemistry, mineralogy, and geology in the same year. The collections were placed on display in the second story of Harvard Hall, where the Mineral Cabinet remained for thirty-three years.[21]

Although Harvard University was the principal center of mineralogical activity in the Boston area in this period, it was not the only one. With the founding of the *New England Journal of Medicine and Surgery* in 1812, a new avenue of publication was opened. The editors of the journal, like their counterparts in Philadelphia and New York, were glad to receive contributions to American natural history. In 1813 they published a long review of Theodric Romeyn Beck's address to the Society for the Promotion of Useful Arts on the state of mining and mineralogy in the United States, adding their own catalog of "minerals and fossils which have been observed in Massachusetts proper, and the District of Maine," subdivided into "Earthy fossils," "Inflammable fossils," and "Metals." To Beck's list of mining and related manufactures they added a steel wire plant at Medford, Massachusetts, a manufactory of copperas [ferrous sulphate] at Winthrop, Maine, one of sulphuric acid at Charlestown, and one of carbonate of magnesia near New Bedford, "where works sufficiently extensive have been erected to produce from fifteen to twenty thousand pounds annually." At the same time they noted that the inhibition placed on American commerce by Jefferson's embargo, Madison's non-intercourse acts, and the ensuing War of 1812 had produced a "metallurgic mania" in Massachusetts during which many businessmen had made ruinous investments in mining and metallurgical ventures without adequate information about the quantity and nature of the mineral ores or the competence of the artisans

[17] Levi Hedge to Parker Cleaveland, Cambridge, March 3, 1815, Cleaveland Correspondence, Bowdoin College Library.

[18] See below, p. 89.

[19] John W. Webster to Benjamin Silliman, Boston, November 11, 1819, Silliman Correspondence, Beinecke Library, Yale University. On November 14 of the same year Webster wrote Silliman that he was asking $10,000 for his collection of 20,000 specimens. To Parker Cleaveland in a letter dated Boston, October 17, 1818 (Cleaveland Correspondence, New York Public Library), Webster wrote: "When I went to Europe I took out all my American specs & recd others from home while there, by which my collection was much increased & contains now I should say from 4 to 6,000 specs."

[20] John Gorham to Benjamin Silliman, Boston, March 18, 1817, Silliman Correspondence, Beinecke Library, Yale University.

[21] Harvard Memorial Society, "History of the Mineralogical Collection," in: *Official Guide to Harvard University* (Cambridge, Mass., 1907), cited in Meisel, *Bibliography of American Natural History* 2: p. 63. This account adds: "It [the collection] increased slowly, and about 1840 contained 26,000 specimens, including rocks and other miscellaneous material. It owes its present value [1907], both in quality and size, chiefly to the late Josiah P. Cooke, Erving Professor of Chemistry and Mineralogy from 1850 to 1894."

employed in these enterprises. Despite these reverses, the editors observed, "the face of the country has in part been explored, . . . many beds of ores and a multitude of articles capable of being manufactured have been discovered, and . . . those in whose possessions they now are, will profit by the errors of their predecessors, and render their processes not only less complicated and expensive, but also more productive."

As this branch of natural history is more studied, as the art of analysis is acquired, as new roads are formed, quarries and canals are opened, and the country is cleared and cultivated, our fossils and minerals will be brought into view, and we may probably be astonished at the profusion with which the bounteous hand of Providence has scattered those mineral riches which are essential to the comfort, the refinement or the luxury of man.[22]

This review of Beck's account of American mining and mineralogy stimulated Benjamin Waterhouse's son, Dr. John Fothergill Waterhouse, to publish a "Description of Certain Minerals, Found in the State of Massachusetts" as a modest contribution to an eventual synopsis of the minerals of America. "We would wish to know the varieties, both in form and colour, of every species, and its locality [Waterhouse explained] : but that is not all, its geological relations; and we earnestly desire that what is known should be concentrated." The list of minerals was not long, but it was apparent from the author's remarks that he was familiar with the Abbé Haüy's crystallographic approach to mineralogy. Specimens were mostly from Massachusetts, Maine, and Connecticut, but a few were from the vicinity of Philadelphia and Trenton. In all Waterhouse listed seven varieties of feldspar, including under the rubric "compact felspar" Godon's "argilloid" [argillite] as a form of feldspar "changed by decomposition and exposure to the air, and soaking in water with decomposing vegetables, from which the potash might have been derived." "It is compact, hard, perhaps tough, as it may be bruised by the hammer ; no brilliancy; melts before the blow pipe; fracture not conchoidal, easily decomposed: These characters distinguish it from Chert or Horn-stone."[23] Unfortunately, young Waterhouse's career was soon cut short. In 1813, the year in which the first part of his "Description of Minerals" was published, he took his M.D. at the University of Pennsylvania and settled in Philadelphia, where he became active in the newly formed Academy of Natural Sciences. Unfortunately, he developed tuberculosis and died in May, 1817, at the age of twenty-six. "Where are now the fruits of

his learning, his rare talents and his matchless industry !" wrote his grieving father.[24]

Besides the lists of mineral localities it published, the chief contribution of the new medical journal to American mineralogy was a long essay review of six publications on the mineral waters of Saratoga and Ballston in the state of New York. These springs had come to the attention of Sir William Johnson, superintendent of the Iroquois tribes for the British Crown, in 1767 when the Mohawk Indians took him there to relieve his suffering from pains of gout. Sir William benefited from them so much that he reserved one of the wells, the so-called Public Well, in Ballston for public use in the original grant of the adjoining lands to private individuals. After the Revolutionary War the springs at both villages were developed commercially and soon became resorts not only for fashionable people like Colonel Gibbs and Robert Gilmor, Jr., but also for poor invalids, "who flock here from all parts of the United States, and are solely dependent upon private munificence for their subsistence." Similar developments took place at Schooley's Mountain in New Jersey, White Sulphur Springs and Warm Springs in Virginia, and many other less well-known spas.[25]

In time American physicians, chemists, and mineralogists made chemical analyses of these various mineral waters, partly from scientific curiosity but more to assist physicians in prescribing the waters for various disorders. The medicinal properties of the waters at Saratoga had been described by a Dr. Constable of Schenectady in 1770. In 1793 Dr. Samuel Tenney sent a notice of them to the *Memoirs* of the American Academy of Arts and Sciences, and Dr. Valentine Seaman published in New York his *Dissertation on the Mineral Waters of Saratoga, Including an Account of the Waters of Ballston.* In 1808 Dr. David Hosack reported a French chemist's analysis of the Public Well at Ballston in Mitchill's *Medical Repository.* Dr. John Griscom's qualitative analysis of the Ballston and Saratoga springs appeared in Bruce's *American Mineralogical Journal* three years later. Griscom confessed his perplexity at the presence of calcium and magnesium compounds and free carbon dioxide in these waters in view of the absence of calcareous and magnesian rocks in the vicinity of the springs. He could only conjecture with Dr. Seaman that the carbon dioxide was evolved "by the powerful action of heat upon calcareous matter situated far beneath the surface." It was not until many years later that drilling operations to increase the flow of the springs revealed a fault in the Little Falls dolomite underlying the strata of shale and limestone as the principal cause of

---

[22] Review of T. R. Beck, *Annual Address . . . before the Society for the Promotion of Useful Arts . . . on the 3d of February, 1813* (Albany, 1813), *New England Jour. Medicine and Surgery* 2 (1813) : pp. 184–185.

[23] John Fothergill Waterhouse, "Description of Certain Minerals, Found in the State of Massachusetts," *New England Jour. Medicine and Surgery* 2 (1813) : pp. 261–264; 3 (1814) : pp. 126–128. The quotation is from p. 128, note.

[24] "Notice of the Late Dr. Waterhouse," *Jour. Acad. of Nat. Sciences Philadelphia* 1 (1817) : pp. 31–32.

[25] See Harry B. Weiss and Howard R. Kemble, *They Took to the Waters* (Trenton, New Jersey, 1962), pp. 208–212.

TABLE 1

BALLSTON
1. Public Well

	M. Soda	Carb. Lime	Carb. Mag.	Oxid. Iron	Carb. Acid
Meade	72	16	20	1.73	105
Steel	68	32	1.08	3.03	90
French	65	46		8.44	300

2. Low's Well

	M. Soda	Carb. Lime	Carb. Mag.	Oxid. Iron	Carb. Acid
Meade	74	23	12	1.73	104
Steel	61	27	0.64	2.59	95

SARATOGA
1. Congress Well

	M. Soda	Carb. Lime	Carb. Mag.	Oxid. Iron	Carb. Acid
Meade	178	47	29	0.86	114
Steel	204	77	1.45	2.67	148
Dana	184	62	12		

2. Round Rock

	M. Soda	Carb. Lime	Carb. Mag.	Oxid. Iron	Carb. Acid
Meade	71	26	17	0.43	114
Steel	84	46	0.64	3.24	109
Seaman	59	65		2.94	69

the many mineral springs in the vicinity of Ballston and Saratoga. The fault runs from northeast to southwest through the town of Saratoga Springs. No spring has been discovered to the west of the fault, and none is known very far east of it. Of the various explanations offered to account for the free carbon dioxide emission from the springs, the most acceptable attributes it to expiring volcanic or igneous action.[26]

In 1817 Dr. William Meade of Philadelphia and Dr. John Steel of Saratoga Springs, each working separately, published the results of qualitative and quantitative analyses of water from several of the springs at Ballston and Saratoga, together with comments on the medical uses of the waters (table 1). Their papers, together with those of Griscom, Seaman, an anonymous French chemist, and J. F. Dana's analysis of the Congress Spring at Saratoga were reviewed in the *New England Journal of Medicine and Surgery* in October, 1817.

The analytical procedures employed by all of the investigators were essentially those recommended by Thomas Thomson of the University of Glasgow in Scotland, but some of the investigators reported the

solid and gaseous contents of the water by the pint, others by the quart or gallon. The reviewer reduced all of these reports to the common measure of grains of solid constituents per 100 cubic inches and of cubic inches of gases per 100 cubic inches to facilitate comparison of the analyses. Even at the qualitative level, the reviewer noted, the analyses showed wide discrepancies:

A single glance at the results of these analyses [of the Ballston spring water], is sufficient to show that they differ exceedingly from each other, not only in relation to the salts, which are the products of evaporation, but to the relative proportions in which they exist. Of the muriates, Dr. Meade finds those of soda, lime, and magnesia, while Dr. Steel denies that the waters contain any other than common salt; on the contrary, the latter discovers carbonate of soda, which could not be detected by the former. The French chemist obtains two muriates and two carbonates. . . .[27]

[26] For a twentieth century view of these problems, see James F. Kemp, "The Mineral Springs of Saratoga," *New York State Museum Bull.* 159 (Albany, N.Y., 1912): p. 6.

[27] Review of six publications on the mineral waters of Saratoga and Ballston, New York, *New England Jour. Medicine and Surgery* 6 (October, 1817): p. 372. The publications reviewed were: Valentine Seaman, *A Dissertation on the Mineral Waters of Saratoga, Including an Account of the Waters of Ballston* (New York, 1809); "Analysis of Ballston Water, Communicated by Dr. Hosack to Dr. Miller," *Medical Repository* 5, 2nd Hexade (1808): pp. 214–215; John Griscom, "Observations and Experiments on Several Mineral Waters,

Quantitatively, the discrepancies were equally disturbing:

If we now turn our attention to the numbers which represent the proportions in which these saline bodies are present, we shall find them equally at variance. . . . It is difficult to account for the very great differences observed. . . . Very considerable errors must exist somewhere; for the amount of the whole saline mass obtained by evaporation, does not vary in any great degree, and it is therefore improbable that the strength of the waters [of the Congress Spring at Saratoga] should, at different times, fluctuate to such an extent as to afford Dr. Steel 77 grains of carbonate of lime, and Dr. Meade only 47 grains, or that the latter should find 29 of carbonate of magnesia, where but one grain and a half were obtained by the former. We can more readily believe that the two carbonates might have been confounded. . . .[28]

The situation was indeed confusing. These early chemists did not detect barium, strontium, potassium, bromine, manganese, fluorine, lithium, and boron in the mineral waters at Saratoga, and their separation, particularly of calcium and magnesium, was incomplete. The reviewer was justified in concluding that "it would puzzle the physician, who sends his patients to the springs to decide, whether *he* prescribes muriates or carbonates, or *they* are to drink any thing more than a solution of common salt, chalk, magnesia, a little iron, and a great deal of carbonic acid."

One wonders whether the author of this learned and trenchant review was Dr. John Gorham, Erving Professor of Chemistry and Mineralogy at Harvard College. Gorham was one of the editors of the *New England Journal of Medicine and Surgery* and an enthusiastic mineralogist. Certainly he was better qualified than any of his colleagues to write such a review.

About the same time that the *New England Journal of Medicine and Surgery* was launched, efforts were under way to establish a natural history society in Boston. A group was formed in December, 1814, with Judge Davis as president, Dr. Jacob Bigelow as Corresponding Secretary, and Dr. John W. Webster as Cabinet Keeper. In 1815 the society adopted the name Linnaean Society of New England; in 1820 it was incorporated by the Massachusetts legislature. According to a notice published in the *New England Journal of Medicine and Surgery* in 1816, the Society had a museum of natural history occupying "a spacious hall over the new south market house."

Among the quadrupeds may be mentioned the Lion, Tiger, Leopard, Catamount, Wolf, Bear, Stag, Sea-Elephant, and a great number of smaller species, principally native. The birds amount to nearly three hundred in number, and consist of elegant species in every order, and of every size, from the Albatross and the large Sea-Eagle of North America, to the minutest Humming birds of Cayenne. A majority of the birds are natives of this country, but there are many beautiful species from tropical countries. . . . The fishes are prepared in uniform half specimens fixed upon a white ground, and afford a fine display, being sufficiently numerous to nearly cover one end of the hall. The insects and shells amounting to some thousands in number, include many rare and elegant species, both native and foreign, and among them may be mentioned, a fine collection of insects from China, and of shells from the Isle of France and Calcutta. The mineralogical specimens already fill four large cabinets, and the herbarium of native plants, yet in its infancy, is perhaps inferior to none for the neatness, and perfection of its specimens.

The whole collection, with the exception of the fishes and a few other specimens, have been inclosed at a great expence, in mahogany cases with glass fronts. They have been arranged according to their orders and genera, as far as the progressive nature of the collection would admit; and most of the specimens are labelled with their scientific and English names. The labour of investigation and arrangement has been performed by the different members of the society, to whom the several departments have been allotted. The specimens have been prepared by members of the society, and by an artist employed for the purpose; or acquired by fortunate purchases and donations. . . . Their collection is placed on the most liberal footing, being always accessible to persons properly introduced, and being opened once a week for the free reception of citizens and strangers.[29]

Despite its favorable beginning, the Society did not flourish. Its collections grew rapidly, but the problem of providing for their housing and care proved insuperable. Not until the founding of the Boston Society of Natural History in 1830 did Boston acquire a counterpart to the Academy of Natural Sciences of Philadelphia and the Lyceum of Natural History of New York.

One substantial contribution to American mineralogy and geology did emerge from the Linnaean

---

in the State of New York," *Amer. Mineral. Jour.* 1 (1814): pp. 156–163; J. F. Dana, "Chemical Examination of the Water of the Congress Spring, Saratoga," *New England Jour. Medicine and Surgery* 6 (1817): p. 19 ff.; William Meade, *An Experimental Enquiry into the Chemical Properties and Medicinal Qualities of the Principal Mineral Waters of Ballston and Saratoga, in the State of New-York* . . . (Philadelphia, 1817); John H. Steel, *Analysis of the Mineral Waters of Ballston and Saratoga* . . . (Albany, 1817).

[28] *Ibid.*, pp. 372–374.

[29] "Museum of Natural History," *New England Jour. Medicine and Surgery* 5 (1816): pp. 189–191; see also 9 (1820): pp. 311–313. For further information about the Linnaean Society of New England see Meisel, *Bibliography of American Natural History* 2: p. 222 ff.; Thomas T. Bouvé, *Historical Sketch of the Boston Society of Natural History; With a Notice of the Linnaean Society, Which Preceded It (Anniversary Memoirs of the Boston Society of Natural History . . . 1830–1880)* (Boston, 1880), pp. 4–14. In 1823 the Linnaean Society voted to turn its collections over to Harvard College and the Board of Visitors of the Massachusetts Professorship of Natural History, who jointly agreed to erect a building for the collection and to give the members of the Society free access to the collection and to the Botanic Garden. The College failed to keep its part of the bargain, and the specimens mostly went to ruin. What was left of them was transferred to the Boston Society of Natural History shortly after it was organized.

Society's activities: the memoir "Outlines of the Mineralogy and Geology of Boston and Its Vicinity, With a Geological Map" by James Freeman Dana and Samuel Luther Dana, printed as a booklet in 1818 and republished in 1821 in the *Memoirs* of the American Academy of Arts and Sciences. The Dana brothers, from Exeter, New Hampshire, had come to Harvard College in 1813. The younger of the two, James, became John Gorham's assistant, finished his medical degree in 1817, and received an appointment at Dartmouth College in the same year. His brother fought in the War of 1812, took his M.D. at Harvard in 1818, and moved to Waltham, Massachusetts, where he eventually abandoned medicine for applied chemistry, becoming one of the first important industrial chemists in the United States.

The Danas' account of the mineralogy and geology of Boston and vicinity was a distinct improvement on Silvain Godon's sketch, published a decade earlier. The minerals of the region were grouped under four classes; earthy fossils, saline fossils, inflammable substances, and metallic fossils. The first class had two orders: acidiferous and nonacidiferous substances. The single genus of the first order was lime, which included two species, the phosphate and the carbonate. The second order contained only species, nineteen in all, among them quartz, mica, feldspar, basalt, garnet, epidote, asbestos, steatite, chlorite, argillaceous slate, and clay. The second class contained only one species, the sulfate of iron. Hydrogen gas and peat were the only species of the class of inflammable substances. Ores of copper, iron, lead, and manganese comprised the species of metallic fossils. Although it is evident from their discussion that the Danas understood the difference between rocks and minerals, they followed the practice of their time by including rocks, soils, and gases among the minerals in their classification.

For each of the "minerals" in their table the Danas gave a concise description and locality. They noted, for example, that large crystals of garnet were embedded in the granite at Bedford and that smaller crystals and "rounded masses" of this mineral occurred in the granite at Cambridge, West Cambridge, and Charleston. Basalt, they wrote, occurred in "beds" in argillite at Charlestown and in "rounded masses" at Cambridge and Charlestown. The true nature of this rock they had verified by comparing it with specimens from the Giant's Causeway in Ireland. Their description of "greywacke" (now called conglomerate) was good if one substitutes "pebbles" for "nodules" and "boulders" for "rolled masses."

The Danas also included a section on geology. Here they faithfully followed Werner, dividing rocks into primitive, transition, and alluvial classes. The primitive rocks were subdivided into granite, argillite, primitive trap, porphyry, and syenite; the transition rocks into amygdaloid [amygdaloidal melaphyre—a local Boston term for altered basalt and andesite] and greywacke [conglomerate]; the alluvial deposits into sand, pebbles, clay, and peat. As a finishing touch, they indicated the surface locations of these rocks in color on a remarkably accurate map of Boston and its vicinity extending from Danvers in the north to Milton in the south and Weston in the west. All in all, the Danas had provided a detailed and comprehensive introduction to the geology and mineralogy of the Boston area.[30]

Apart from the Danas, the most active mineralogist and geologist of the Linnaean Society of New England was its secretary, Dr. John W. Webster. In the winter of 1818–1819 Webster gave a course of lectures by public subscription, but it appears to have been poorly subscribed. "When I have closed the *only* course of lectures I ever mean to give this unscientific people," Webster wrote to Parker Cleaveland,

it will be my first object to arrange my cabinet & shall lay aside whatever you may wish from my duplicates. I say this unscientific people, for but four tickets for my lectures have been taken; & although I have the most ardent desire to devote myself to chemistry and mineralogy, fear I must abandon both & think of naught but strictly professional [i.e., medical] subjects.[31]

Webster was a competent mineralogist, as shown by his account of the meteorite which fell in Nobleboro, Maine, in 1823 and his analysis of sea water from Boston Harbor in 1825.[32] He taught chemistry, mineralogy, and geology at Harvard from 1824 until his death, but, far from making a reputation as the Benjamin Silliman of Harvard, he is remembered chiefly as the perpetrator of an atrocious murder, for which he was hanged in 1850. Not until the founding of the Boston Society of Natural History in 1830 did the Cambridge-Boston area begin to emerge as a center of significant research in the earth sciences.

---

[30] James F. Dana and Samuel L. Dana, "Outlines of the Mineralogy and Geology of Boston and Its Vicinity, with a Geological Map," *Memoirs Amer. Acad. Arts and Sciences* 4 (1821) : pp. 129–223. Originally published in 1818 by the Boston firm of Cummings and Hilliard. The authors are greatly indebted to Professor Marland P. Billings, emeritus professor of geology, Harvard University, for providing them with a critique of the publications of Godon and the Danas on the mineralogy and geology of the Boston area.

[31] John W. Webster to Parker Cleaveland, Boston, October 17, 1818, Cleaveland Correspondence, New York Public Library.

[32] John W. Webster, "Chemical Examination of a Fragment of a Meteor Which Fell in Maine, August 1823, and of Green Feldspar from Beverly, Mass.," *Boston Jour. Philosophy and the Arts* 1 (1824) : pp. 386–390; John W. Webster, "Chemical Analysis of the Sea-Water of Boston Harbor," *ibid.* 2 (1825) : pp. 96–97. In a letter to the authors, Clifford Frondel describes these two chemical analyses as "creditable," adding: "the composition given for the meteorite, with 18.3 per cent sulfur, looks absurd but only because he [Webster] chanced to include a fragment of the iron sulfide troilite."

VI. PARKER CLEAVELAND AND THE UNIFICATION OF AMERICAN MINERALOGY

While mineralogy was struggling to gain a secure foothold in Cambridge and Boston, a young Harvard graduate named Parker Cleaveland was establishing it firmly at Bowdoin College. The son of a physician in Byfield, Massachusetts, Cleaveland graduated from Harvard in 1799, taught briefly at Haverill and York in the Maine district, and was then appointed instructor in mathematics and natural philosophy at Harvard. In 1805 he accepted a similar appointment at Bowdoin, a small college established in Brunswick in 1794. During his first two years at Bowdoin, Cleaveland spent most of his time on natural philosophy, exchanging observations with Benjamin Vaughan of Hallowell. Vaughan was one of several sons of Samuel Vaughan, an English merchant who emigrated to Philadelphia in 1782 and played a prominent role in revitalizing the American Philosophical Society, in planning and raising money for Philosophical Hall, and in landscaping the State House gardens. Another son, John, remained in Philadelphia after his father's return to England, serving as treasurer and librarian of the American Philosophical Society until his death in 1841. A third son, Samuel, settled in the West Indies, where he devoted the profits from his commercial ventures to amassing a splendid apparatus and library for the study of mineralogy. Still another son, William, settled in England, where he was in close touch with British men of science. Meanwhile Benjamin Vaughan, after an interesting career as a diplomatic agent, retired to the family estate in Maine to lead the life of a gentleman scientist. He soon became a leading benefactor of Bowdoin College, contributing generously to its library, scientific apparatus, and collections and encouraging Parker Cleaveland in his efforts to improve and develop the study of the sciences at Bowdoin.[1]

In 1807 a chance event turned Cleaveland's scientific interests in the direction of mineralogy. During the course of excavations to facilitate the floating of lumber downstream from one of the mills on the Androscoggin River, the workmen came across some pieces of quartz and iron pyrites which they mistook for diamonds and gold. When they brought these treasures to the professor of mathematics and natural philosophy to confirm their hopes, Cleaveland had to admit that he could not say with any certainty what the substances were. His interest was aroused, however, and he began collecting minerals, preparing lectures on chemistry and mineralogy, and enlisting Benjamin Vaughan's aid in starting a cabinet of minerals at Bowdoin and acquiring the latest min-

eralogical works for the library. In May, 1808, Vaughan wrote to Cleaveland:

I am glad that the work of a cabinet of minerals at Brunswick is begun. To him that hath, shall more be given. If I see anything which may be useful, I will send it. But when our country people find any thing, it is to be sure gold or silver or something calling for secrecy, & they do not tell *where* the article produced was found. . . .

Dr. Goreham [*sic*], who is going to read chemical and mineralogical . . . lectures in Boston, will be glad for some of the crumbs from your riches in the mineral way. And whatever you obtain *without cost*, if put into my hands, shall go to him. . . .

I sent the second edition of Kirwan by design, as being better than the first. You will find M. Godon recommending it in the anthology [*Monthly Anthology and Boston Review*] for March or April; which is something in its favor. I know Mr. Kirwan well. He is anxious & laborious; has the command of good cabinets; has long toiled in the pursuit; has seen many foreigners to have their opinions; and reads the latest of everything. If there is a third edition, you ought to have it. I am desirous you should get this work, & the new edition of Thomson's chemistry, for the College.[2]

In August of the same year Vaughan sent further encouragement:

I shall send you a few specimens of minerals, but I expect many more.

I am in hopes to persuade a friend to undertake to engage Mr. Bowdoin to make you a present of a collection of minerals bought very cheaply through Mr. Godon for $100. With your contributions to the college, & others through Maine, added to the above, the College will begin to be noted for its collection; & these things always increase, where there is a beginning made, and active friends. I begin to think that Maine may be celebrated for its minerals in future times.[3]

Vaughan's efforts in behalf of the mineral cabinet and the library at Bowdoin continued through the years. In April, 1819, he wrote Cleaveland that he had "got all the English books you desired, & all the French as far as they existed," as well as "journals &c from England, for the subscription fund & college." In October he promised to order Berzelius's *Nouveau système de minéralogie* from Paris for Cleaveland and reported that he had asked Judge John Davis to procure Baron Héron de Villefosse's great treatise *De la richesse minérale* for Harvard College "that you may read it." Vaughan's brother John was commissioned to procure Haüy's *Traité des caractères physiques des pierres précieuses*. The following summer Vaughan informed Cleaveland that his brother Samuel in Jamaica had shipped his library of a thousand volumes, chiefly chemical and mineralogical, to Maine. "I have not heard that he has sent his minerals *in*

---

[1] Sarah P. Stetson, "The Philadelphia Sojourn of Samuel Vaughan," *Penna. Mag. History and Biography* 73 (1949): pp. 459–474.

[2] Benjamin Vaughan to Parker Cleaveland, Hallowell, Maine, May 26, 1808, Cleaveland Correspondence, Bowdoin College Library.

[3] Benjamin Vaughan to Cleaveland, Hallowell, August 26, 1808, Cleaveland Correspondence, Bowdoin College Library.

toto," Vaughan added, "but many are here already." [4]

Cleaveland was not dependent solely on Benjamin Vaughan for assistance in acquiring books, specimens, and descriptions of American mineral localities. In 1808 he met Silvain Godon and William Maclure, who visited him at Bowdoin. In November, Maclure sent him a description of the minerals and rocks he had observed and collected in New England since leaving Brunswick. From Philadelphia two months later Maclure acknowledged the assistance Cleaveland had given him in shipping specimens from Maine and added:

On receipt of it [Cleaveland's letter] I searched the town for Thompsons [*sic*] Chemistry and Kirwan's mineralogy but could find neither. I have been more fortunate at New York where I procured at the rate of 5F per volume Thompsons latest edition in 5 volumes which will be shipt aboard the Harmony Capn Say for Boston care of Mr. Snow as you direct. In the same care shall be forwarded 38 specimens of minerals mostly from Derbyshire in England. . . . I would recommend Jemisons [*sic*] mineralogy in place of Kirwans. As I forwarded 2 copies to Mr. Bowdin [*sic*] and his friend Mr. Winthrop when I was at New York I have taken the liberty to inform Mr. Bowdin that you are much in want of some treatise on that subject and its probable he may spare you his copy as it can be replaced by the first importations. I found the Emerald in a granite near Singsing State of New York and small Christals laminated [?] in Mr. Hosack's Garden 3 miles from New York. . . . I spoke to Mr. Gibs [*sic*] of Rhode Island Mr. Silliman of New Haven & Dr. Bruce of New York who all promised to send you some specimens of foreign minerals which I hope they have done. [5]

Thanks to Maclure's help, Cleaveland was soon exchanging letters and specimens with Bruce, Silliman, and Gibbs. In April, 1809, Cleaveland acknowledged a letter from Gibbs and promised to send him specimens of Maine minerals as soon as the weather permitted him to replenish his supply of duplicates.

My collection is yet quite small [Cleaveland added]. It consists principally of specimens which I have myself collected, during hours of leisure from collegiate duties. A few friends have given me some assistance. It is with much satisfaction, that I receive the intelligence, that you are about to forward me a box of specimens; especially as I am at the time engaged in a course of lectures upon mineralogy. I would here take the liberty of observing that many specimens, which might be too *small* or too *rough* for the cabinet of the amateur, would answer a valuable purpose to a lecturer on mineralogy. [6]

In the following month Cleaveland acknowledged receipt of a "rich feast" of minerals from Gibbs and announced that he was sending in return a box containing "feldspar, gneiss, amphibolic rocks, mica, granites, hornblende, carbonate of lime, quartz, emerald, tourmalin, magnetic iron ore, molybdena, pyrites &c about 37 specimens in all."

Most of them are aggregates, which I suppose corresponds with your geological views. . . . Please to write me, if the size of the larger ones suits your views; also be kind enough to note any *defects* in my *description;* for, Sir, I never saw a cabinet; all my knowledge of mineralogy is obtained from authors and the study of nature. I have made one or two queries in my labels. All the specimens, which I have sent you were collected within 4 miles of Bowdoin College, one excepted. [7]

Gibbs also visited Cleaveland in Brunswick, and Cleaveland wrote in September that he was hoping for a second visit from him. Meanwhile Cleaveland continued to press Gibbs for specimens for the Bowdoin College collection. "The situation of this College is favorable for the study of mineralogy," he wrote, "and its Trustees & overseers have discovered a strong partiality to patronize mineralogical pursuits. . . . I shall say no more, than to request that, when you open your collection, you will permit the Institution of Bowdoin College to share as largely in those specimens, which are not wanted for your own cabinet, as you think it deserves." [8]   In 1812 Cleaveland announced the acquisition of James Bowdoin's "fossils" and crystallographic models:

The Crystallography is exceeding valuable [he wrote]. There are about 60 pieces, made very accurately, of porcelain clay, exhibiting primitive forms, and the structure of secondary crystals, according to M. Haüy's theory; also about 330 models of crystals, made of clay & painted; of good size to be employed at lecture, and finely exhibiting the various truncations, bevillings [*sic*] &c. This collection is invaluable to me in the business of instruction.
The collection of fossils contains about 450 specimens, including some duplicates. The cursory view, I have taken, discovers many fine specimens among the number. It is undoubtedly a valuable collection, and with those, with which you are about to furnish our college, will enable us to afford our pupils very great facilities for the study of mineralogy. [9]

Gibbs's friend Archibald Bruce was also in contact with Cleaveland. In October, 1809, Bruce responded

[4] Benjamin Vaughan to Parker Cleaveland, April 30, 1819; October 22, 1819; August 6, 1820, Cleaveland Correspondence, Bowdoin College Library. On October 31, 1820, Vaughan reported receipt of 19 boxes from his brother in Jamaica.

[5] William Maclure to Parker Cleaveland, Boston, November 3, 1808; Maclure to Cleaveland, Philadelphia, January 10, 1809, Cleaveland Correspondence, Bowdoin College, Library. The "emerald" referred to was probably green beryl or aquamarine.

[6] Parker Cleaveland to George Gibbs, Brunswick, Maine, April 18, 1809, Gibbs Correspondence, Newport Historical Society. On July 14 Cleaveland wrote: "I am anxious to see your geological account of Rhode Island."

[7] Cleaveland to Gibbs, Brunswick, Maine, May 15, 1809, Gibbs Correspondence, Newport Historical Society.

[8] Cleaveland to Gibbs, Brunswick, Maine, November 6, 1809, Gibbs Correspondence, Newport Historical Society. In the same letter Cleaveland writes that he has heard from Adam Seybert: "I am under obligations to yourself for the introductory letter."

[9] Cleaveland to Gibbs, Brunswick, Maine, February 26, 1812, Gibbs Correspondence, Newport Historical Society.

to a gift of Maine minerals from Cleaveland with a collection of foreign specimens:

I have forwarded to you to the care of Major [Joseph] Tilden of Boston a collection chiefly European [Bruce wrote]. I regret they are not all of a cabinet size, tho from circumstances of their rarity, size cannot always be commanded. I have ticketed them individually so that I trust you will find no difficulty in arranging them. Among others you will find some single crystals, which are only to be valued on account of their locality such as the Peruvian [Colombian] emerald, the Brazilian & Saxon [?] topaze, the Ceylon & French hyacinth, &c. These in my cabinet I have fixed on small pedestals of wax as the best mode of displaying them. I have sent some Peruvian specimens, as the pyroxene, idocrase &c. & I trust I shall be enabled in the course of a short time to add a few specimens of the products of Etna having yesterday received accounts of a collection on the way from Sicily.[10]

In Benjamin Silliman, Cleaveland found a correspondent who was to be of enormous help to him in all his chemical and mineralogical investigations. In May, 1811, Silliman responded to Cleaveland's present of Maine minerals with a box of specimens, declaring himself "very much gratified to find that mineralogy is cultivated so industriously & with such just discrimination & correctness in a part of our country which (judging from the specimens) appears to be particularly interesting:"

The specimens which I send you in return are chiefly from this vicinity [Silliman added] & are principally of my own collecting with occasional aids from my pupils. I send you a few foreign specimens; I would gladly have augmented the number, but, generally speaking, our foreign specimens are so arranged in connection with others that few of them can be withdrawn without breaking the collection. I should, however, be less influenced by this motive were they my own property; they belong to the Institution. I shall be very happy to continue an exchange of minerals with you whenever mutually convenient, and it will give me pleasure to become personally acquainted with you.[11]

Three years passed before Cleaveland could take advantage of Silliman's invitation to visit him in New Haven. By that time Colonel Gibbs had deposited his collection on loan at Yale, giving Cleaveland an added incentive to make the journey. Arriving in May, 1814, he was cordially greeted by Silliman. No account of their conversations has been recorded, but the two men must have found each other extremely congenial. Both were ardent scientists, both Federalist in their political views, and both devout Christians. One can imagine the pleasant hours they spent together examining the treasures of the

mineral collection, discussing plans for Cleaveland's textbook, lamenting the war with Great Britain, and comparing notes on academic and religious matters at Bowdoin and Yale. Colonel Gibbs did not arrive until after Cleaveland had departed, but he visited Cleaveland in Brunswick the following year, and the two men improved their acquaintance while ranging the countryside in search of minerals.[12]

By this time Cleaveland was hard at work on a textbook of mineralogy and geology suited to the needs of American readers. In the same letter in which he explained to Silliman that Colonel Gibbs seemed to have fallen out with some of the Boston mineralogists he commented that he was thinking of editing an American edition of the mineralogical section of Thomas Thomson's chemical textbook for the Boston publisher Cummings and Hilliard. The advantages and disadvantages for American students of the various European works on mineralogy had been outlined for him by John Gorham in 1809 in response to Cleaveland's letter of inquiry:

I know of no small modern work on mineralogy which might answer as a text book for a course of lectures on that subject. The work of Murray you mention is confined to geology and is meant as an answer to Playfair's Illustrations of the Huttonian Theory. A small manual on this science is much wanted and I think that portions of Thompsons [sic] Chemistry relating to it might conveniently answer, if separately printed.

The best works in English on Mineralogy are Kirwans, which is now almost out of date and Jamesons lately published in 3 volumes. To the latter however there are some objections. Accum has lately published a small but valuable work on the analysis of minerals. The 3d of Murrays chemistry contains an excellent though rather concise account of the native combinations of the earths and metals.

The foreign works, particularly those of the French are superior to the English. Brochant in 2 vols 8vo affords a very correct view of the system of Werner. I think it very valuable. Brogniart [sic] has also lately published a system which has a high reputation. The French mineralogists, however, in general prefer the work of the Abbe Hauy, but for us the other systems are preferable as they contain less number of new names and are sufficiently appropriate in their nomenclature. The new dictionaries also of chemistry & mineralogy of Nicholson & the Aikens, each 2 vols. 4to give a good view of the more recent discourses & analysis in this science.

These I believe are all the modern works on this

[10] Archibald Bruce to Parker Cleaveland, New York, October 12, 1809, Cleaveland Correspondence, New York Public Library.

[11] Benjamin Silliman to Parker Cleaveland, New Haven, May 28, 1811, Cleaveland Correspondence, Bowdoin College Library.

[12] Cleaveland to Gibbs, December 23, 1813 and May 7, 1814, Gibbs Correspondence, Newport Historical Society. Also Cleaveland to Gibbs, March 17, 1815: "I am glad you intend sending me the geological specimens of Rhode Island. It will be useful to me to compare them with your geological description. I have prepared me a large *blank* for beginning a mineralogical map of the precinct of our College, for several miles around. Unless *individuals* contribute on a *small scale* from different parts of our country, we can never generalize our observations respecting a large extent of country." In a P. S. Cleaveland adds: "I fear the Mineralogical Society in Boston will have to be their own Medici."

subject, except Klaproths which is merely an account of the analysis of different minerals.[13]

After discussing the various possibilities with Cummings and Hilliard and inquiring whether Silliman himself intended to publish a guide to the study of mineralogy, Cleaveland decided to write a textbook himself, with special emphasis on the localities of American minerals. He began the work with characteristic energy and enthusiasm, but soon found the task a difficult one. His duties at Bowdoin scarcely permitted him to do field work beyond the sound of the college bell, he wrote Silliman.

Were I not so strongly attached to mineralogy, I should be discouraged in my present undertaking. This will not appear strange, when I observe to you, that my only leisure consists of *intervals* between *3 recitations* and 1 experimental lecture a day. All that I can do is to *rise* almost every hour, *report* a little *progress*, with the hope of soon *sitting* again. In consequence of the low state of the College funds, the Overseers have this year diminished our usual number of tutors. This has devolved on us, who remain, an unusual share of duty—and so much of it on myself, that, unless the prospect brightens, I shall be tempted to give up my present situation.[14]

To make matters worse, the United States had become involved in a war with Great Britain, much to the distress of pro-British Federalists like Cleaveland and Silliman, and the coast of Maine was invaded and ravaged by British troops. Cleaveland described the situation in a letter to Silliman in September, 1814:

There is now one army of nearly two thousand men within seven miles of my house, another of nearly three thousand at the distance of eighteen miles, and another, this larger, about twenty-six miles west of us. It has been supposed, that *Brunswick* is in very considerable danger of an attack, as we have two large manufacturing establishments, and two iron furnaces, one of which is constantly bringing forth the *means of annoyance*, as Mr. Madison calls them, that is, cannon-balls; and more especially, as we are so easily accessible from the sea. I have not, perhaps, felt so much consternation as many of my neighbors, because I have ever believed that college-ground would be held *sacred*. I have, however, for some time kept my most valuable papers, &c., in trunks, ready to decamp when I see contiguous buildings in flames. So much, and all to gratify the cursed democracy of this country. Can brother Day [Jeremiah Day, professor of mathematics and natural philosophy at Yale] keep

cool, even when breathing the sober atmosphere of mathematics? I confess I cannot, and, when I reflect on the present state of our native country, and perceive "Troja fuit" written on all our greatness, my only relief is to sally forth with my hammer, and vent my feelings in the demolition of some rugged cliff of granite that rise[s] on the banks of the Androscoggin. But I have insensibly gotten into the mineral kingdom, and will now endeavor to feel a little calmer, notwithstanding these turbulent times. I still go on, and suffer no day to pass *without a page or two*. For several reasons, however, the work cannot appear before the winter. Indeed, were I this day ready for the press, I should doubt the expediency of proceeding instantly, such is the universal state of excitation and alarm throughout our country. I am yet to receive considerable assistance from two or three gentlemen in Baltimore, which must, of course, be delayed by recent events in that vicinity.[15]

As his reference to the mineralogists of Baltimore shows, Cleaveland extended his network of correspondence far and wide in the hope of assembling a comprehensive list of mineral localities in the United States. His letter to Robert Gilmor, Jr., requesting his assistance in this undertaking arrived just as Gilmor had completed his "Descriptive Catalogue of Minerals Occurring in the Vicinity of Baltimore" for Bruce's *American Mineralogical Journal*. In his reply to Cleaveland, Gilmor expressed satisfaction that Colonel Gibbs had put Cleaveland in correspondence with him and indicated his willingness to exchange minerals. He added that Cleaveland could expect an account of "our minerals" from Dr. Horace Hayden, a Baltimore dentist who had become interested in mineralogy and geology: "he is a very accurate observer, & promises to be one of our most discriminating mineralogists." In the same letter Gilmor described for Cleaveland a Baltimore mineral that he had not yet succeeded in identifying.

It has the singular property of becoming with borax [?] of a most beautiful emerald green while heated, but on cooling this bead is colourless and this can be renewed as often as one wishes. I carried it to New York last summer to show to Dr. Bruce, and on looking over his cabinet I discovered among his *garnets* a massive substance which I suspected to be of the same nature, and on trying it with the blowpipe, we found it possessed the same phosphorescent property. What it can be I can scarcely guess. It is of a dark garnet colour, semitransparent in small fragments, & resinous in aspect, not unlike dark pieces of green copal or amber. It is tolerably heavy & is evidently metallic. Query does it not contain manganese? Can it be a phosphate of manganese? Neither Brongniart, M. Hauy, nor Brochant mention such a substance as this appears to be.[16]

[13] John Gorham to Parker Cleaveland, Boston, October 23, 1809, Cleaveland Correspondence, New York Public Library.

[14] Cleaveland to Silliman, Brunswick, November 13, 1813, Silliman Correspondence, Beinecke Library, Yale University. On June 18, 1813, Cleaveland had written to Silliman concerning Silliman's plan to publish a brief guide for his students: "I hope you will be ready for publication *soon*. No one can feel the need of a suitable work more, than I do; for with us the students are questioned in Mineralogy, as well as Chemistry at daily recitations. You will understand, that my wish to know *when* the Public will be favored with your work arises from a desire to free myself from any obligation to Cummings & Hilliard. They certainly will not think of republishing Thompson."

[15] Cleaveland to Silliman, Brunswick, September 20, 1814, Beinecke Library, Yale University.

[16] Robert Gilmor, Jr. to Parker Cleaveland, Baltimore, March 1, 1814, Cleaveland Correspondence, Bowdoin College Library. Concerning Gilmor's mysterious mineral, Professor Clifford Frondel comments: "I do not know the identity of the mineral. The blowpipe test described does not indicate if an oxidizing or reducing flame was employed (which changes the results) and in any case does not correspond to any par-

The Baltimore mineralogists were not the only ones eager to contribute mineral localities for Cleaveland's treatise. From Middlebury College, Professor Frederick Hall sent specimens and promised to use Cleaveland's book in connection with his mineralogical lectures, "which I annually deliver to the students of our college." His pupils, he added, were too poor to buy an expensive text, but "they need *something*, and, if it were cheap, would procure it, to give them mineralogical information." [17] Anxious to aid in the enterprise, Hall later reported to Cleaveland the localities where he had found flint, marble, coccolite [pyroxene], garnet, rock crystal, porcelain earth, clay slate, mica, soapstone, cyanite, gypsum, and iron and lead ores. From Washington, D.C., Adam Seybert, now a congressman, forwarded Cleaveland's appeal for help to Professor Elisha De Butts at the College of Maryland, recommending also Dr. Horace Hayden in Baltimore and Charles Wister in Philadelphia. From Thomaston, Maine, Joseph Sprague sent specimens from the marble quarries. From Boston, Major Joseph Tilden reported the results of his geological tour in Maine in 1814. From New Braintree, Massachusetts, a former student sent a box of minerals collected by a local physician and a local clergyman, both enthusiasts for mineralogy. From Northampton, Massachusetts, Dr. David Hunt, the physician who first interested Edward Hitchcock in mineralogy and geology, sent many contributions. [18] Gibbs and Bruce were warm in their endorsement of

Cleaveland's project but slow in sending him their mineral localities.

In New Haven, Benjamin Silliman was a tower of strength to Cleaveland, supplying specimens, performing analyses and experiments, advising Cleaveland on the general plan of his book, urging Gibbs and Bruce to contribute information about mineral locations, and answering innumerable detailed questions about specimens and localities. In November, 1814, Silliman gave Cleaveland his views as to the plan of the projected text. There should be an introductory section, giving a general idea of minerals as distinguished from animals and vegetables, defining mineral species, noting the place of mineralogy among the branches of natural history, and describing the classification of minerals as being founded on (1) external characters, (2) the structure and form of crystals, especially the integrant molecule, and (3) chemical composition. The main body of the work should separate minerals into five classes: (1) acids uncombined, (2) salts not metallic, (3) earthy substances, (4) combustibles not metallic, and (5) metals. There should also be an appendix devoted to rocks. Silliman went on to indicate in greater detail what the introductory section should include, what minerals he would include in the first three classes, and what rocks should be described in the appendix. "Beyond this," he concluded, "my arrangement is not sufficiently digested to be worthy of attention." [19]

Cleaveland followed Silliman's general plan, although he altered it in many details. The introductory section, comprising some eighty-five pages, included in rich detail all that Silliman had outlined, although in a different order. Cleaveland did not employ the classification Silliman had recommended, however, nor did he follow his recommendations concerning the geological section.

In May, 1815, Cleaveland wrote Silliman that the work was approaching completion, adding:

I hope you will find leisure to write me in answer to this in the course of a *fortnight*. I much want your opinion on the nomenclature of crystals. I am now engaged in the geological and last chapter of my work. One thing is certain. I have *myself* received very great benefit from the preparation of this work for I have been faithful in the examination of authors, and *specimens*, as far as necessary, or as far as was within my power, I have sometimes devoted one or two days to studying a species before writing its description. After all doubts and difficulties remain, and whether any other person will receive 1/10 of the advantage from the perusal, that I have from the preparation, is very problematical.

What ails Bruce & Gibbs? I cannot get any localities from them yet. Your list of localities shall be returned by the first . . . private opportunity. [20]

---

ticular element. It does not correspond to manganese. The test probably refers to a mixture of elements, chiefly of iron."

[17] Frederick Hall to Parker Cleaveland, Middlebury College, November 15, 1813, Cleaveland Correspondence, Bowdoin College Library. Hall graduated from Dartmouth in 1803, tutored there in 1804–1805, then taught at Middlebury College from 1806 to 1824. He gave a cabinet of minerals to Dartmouth College.

[18] There are several letters to and from Dr. David Hunt in the Gibbs Correspondence at the Newport Historical Society and in the Cleaveland Correspondence at Bowdoin College Library. Colonel Gibbs seems to have assisted him greatly in his mineralogical studies. On June 1, 1813, Hunt wrote to Gibbs:

"I shall return ye Brong. [niart] system of Mineralogy by the first safe conveyance. I have taken great satisfaction in perusing it. When I began to study it, three pages was a days work, now I read thirty with ease. I am passionately fond of the French language. Have you a system of Chemistry in that language you can favor me with the perusal of? . . . I should have returned Brogn. [sic] ere this, had I supposed you had commenced yr mineralogical excursions."

On December 27, 1817 he wrote to ask why he had heard so little from Gibbs recently: ". . . is it not in consequence of your having weaned yrself from our favorite study for agricultural pursuits? my zeal is not abated—it is true that my professional duties have become more arduous & of course I have less time to attend it, however, I have started Mr. Hitchcock, Preceptor of Deerfield Academy (20 Miles above me) who is indefatigable in any branch of natural history he undertakes."

[19] Silliman to Cleaveland, New Haven, November 6, 1814, Cleaveland Correspondence, Bowdoin College Library.

[20] Cleaveland to Silliman, Brunswick, May 16, 1815, Silliman Correspondence, Beinecke Library, Yale University.

Silliman soon replied, giving detailed answers to nineteen questions submitted by Cleaveland. The quality of Silliman's assistance can be seen from the following examples:

2. Bruce's native magnesia is not a carbonate (according to his analysis). The effervescence arises from carbonate of lime in the gangue which is however chiefly common serpentine.

3. I have never seen any asbestos which had not a fibrous structure however compact it might be. . . . I think I have seen the stone which he [Brongniart] describes; it was composed of radiated starry fibres dispersing from many centres & seemed to hold intimate relations with asbestos, actynolite, tremolite & soapstone; it was different from the common hard asbestos. . . .

6. I communicated a paper to Gibbs on the chromate of Iron of Baltimore in which I believe I considered it as an oxid of Chrome combined with an oxid of iron & this I think is Vauquelin's idea in the Journal des Mines. . . . .

9. The Chatham cobalt is arsenical. . . .

10. I am now unable to make the requisite experiments to ascertain whether the Washington Dolomite contains magnesia; I think however there is little doubt of it; boil sulphuric acid on some of it till the acid is saturated or the superfluous acid evaporated; dissolve everything soluble in hot water, filter, evaporate, cool. If any magnesia the soluble part will afford bitter prismatic crystals of Epsom salt.[21]

Silliman promised to prod Gibbs and Bruce again, although without much hope of success. "You cannot get them to go through the labour of writing what they know of localities. Your only chance in my opinion is to go to them & take down from their own mouths the facts." By November, 1815, however, he had talked with Gibbs and was more hopeful that the mineral localities would be sent.

Silliman planned to use Cleaveland's text at Yale. Even before it was published it was awarded prospectively in the prize competition established among Yale students of mineralogy by Colonel Gibbs.

I hope your publishers will not delay your work beyond the specified time [Silliman wrote to Cleaveland]; we want it exceedingly; my private course of lectures (by a new arrangement) will commence today, &, if I may judge from the numbers which are pressing into them the zeal for the science is increasing with us. I will thank you to request the publishers to forward me a copy as early as possible & I should even prefer it in sheets that I may have it bound with blank paper for the insertion of new American localities & other interesting observations.[22]

Silliman was also preparing to give a course of lectures on the chemical analysis of minerals, having determined to master this difficult branch of chemical

knowledge "that I may do something towards analyzing our own native minerals." [32]

Published in Boston in December, 1816, Cleaveland's *Elementary Treatise on Mineralogy and Geology* was soon in use at Yale.

Your book has been now some days in possession of my pupils [Silliman wrote]; my course was so arranged that it was very important to my pupils to have as much as possible the work immediately & at my request it was sent forward in sheets & we have had a sufficient number stitched for present use expecting to add what remains when it comes. . . . My impressions are thus far very favorable to the work, but my opportunities of examining have been as yet very limited. . . . I have a private mineralogical class of over 40 & they have manifested much eagerness to buy your book.[24]

The work was dedicated to Benjamin Vaughan, without whose assistance and encouragement Cleaveland would scarcely have been able to produce it. Vaughan himself first became aware that the book had been dedicated to him when he saw a copy at Cummings and Hilliard on a trip to Boston. His peculiar reaction to sight of the dedication perhaps explains why Cleaveland had not informed him beforehand.

When I first saw, in the hands of Mr. Cummings your bookseller, the dedication of your work on mineralogy [Vaughan wrote from Boston], I felt strongly disposed to request its immediate suppression; but when I reflected that the work itself was pledged for delivery without delay to the public, & recollected certain circumstances which regard the College & which can be best explained when we meet; I resolved to do some violence to my own feelings on the occasion, & not to cause delay to the publication of the work. But as there is still much in the form of the dedication which cannot properly belong to me, I must request on the event of the work's reaching a second edition (which I trust it will do), either that you will have the goodness to withdraw the dedication altogether, or limit it to the mere mention of the name of the person whom you are thus pleased to flatter. I am convinced that you will perceive that however gratifying it may be to me to have my name associated on such an occasion with yours, yet something may be allowed to feelings on my side which began in youth, & have continued to old age without abatement, & are not altogether unknown to many of my friends.[25]

Whatever Vaughan's scruples about Cleaveland's dedication, the book itself was one of which Vaughan

[21] Silliman to Cleaveland, New Haven, May 24, 1815, Cleaveland Correspondence, Bowdoin College Library.

[22] Silliman to Cleaveland, New Haven, November 11, 1815, Cleaveland Correspondence, Bowdoin College Library.

[23] Silliman to Cleaveland, New Haven, March, 1816, Cleaveland Correspondence, Bowdoin College Library.

[24] Silliman to Cleaveland, New Haven, December 16, 1816, Cleaveland Correspondence, Bowdoin College Library.

[25] Benjamin Vaughan to Parker Cleaveland, Boston, January 4, 1817, Cleaveland Correspondence, Bowdoin College Library. Vaughan seems to have felt unworthy to be called "the patron of general literature, and more especially of Natural Science" and included with those who had "devoted a large portion of life to the cultivation and improvement of deeper and more abstruse sciences." In accordance with his wishes, the dedication of the second edition mentioned only Vaughan's name and his membership in various scientific societies.

might justly feel proud. Others to whom Cleaveland acknowledged indebtedness were Benjamin Silliman at Yale, Horace H. Hayden, Elisha De Butts, and Robert Gilmor, Jr., in Baltimore, Charles Wister and Solomon W. Conrad in Philadelphia, Frederick Hall at Dartmouth, and Bruce and Gibbs in New York. Among the European works on which he had drawn he mentioned especially Kirwan's *Elements of Mineralogy*, Haüy's *Traité de minéralogie* and his *Tableau comparatif des résultats de la cristallographie et de l'analyse chimique*, Brochant de Villiers's *Traité élémentaire de minéralogie, suivant les principes du professor Werner*, Robert Jameson's *System of Mineralogy*, J. A. H. Lucas's *Tableau méthodique des espèces minérales*, and Alexandre Brongniart's *Traité élémentaire de minéralogie*. Werner's method, Cleaveland observed, was valuable for its "technical and minutely descriptive language," whereas that of Haüy was "accurate and scientific." Believing, however, that Brongniart had successfully unified these two approaches, he had decided to adopt Brongniart's plan of classification.

Cleaveland's "Introduction" was divided into four chapters, the first devoted to definitions and preliminary observations, the second to the crystallography, external characters, and chemical composition of minerals, the third to classification, and the fourth to nomenclature. Cleaveland's account of Haüy's method, his primary forms, and his laws of decrement was complete and excellent, although he did indulge in some speculations of his own. The integrant molecules, he declared, remained invariable in the same body, but the description of the elementary particles might change with the progress of chemistry. "Possibly the sulphur and iron may both prove to be compounds." Cleaveland listed and defined all of Haüy's new terms but, following the advice of Silliman, did not attempt to translate them. In the second section of this chapter Cleaveland described twenty-seven physical or external characters, among them electricity, which he believed to be of two kinds, positive and negative, or vitreous and resinous. In the third section Cleaveland described the common methods used in the chemical analysis of minerals, including an account of the action of acids and other tests.

Chapter III discussed the principal systems of classifying minerals. Werner's system, Cleaveland thought, was unscientific. His first, or earthy, class contained nine genera, of which seven were determined by the predominant or characteristic earth they contained. These were zirconian, silaceous, aluminous, magnesian, calcareous, barytic, and strontian. But there were many anomalies. "Thus sapphire is placed in the silaceous genus, although it is nearly pure alumine; and opal, which, in some varieties, does not contain a particle of alumine, is nevertheless referred to the aluminous genus."

For the fact appears to be this; a certain number of external characters, which silaceous minerals usually exhibit, being assumed as generic characters, or as a type of the genus; every mineral, possessing those characters, whether it contain any silex or not, is arranged under the silaceous genus.[26]

Cleaveland conceded that the Wernerian method had its advantages, but chemistry offered a firmer basis for the scientific classification of minerals than external characters.

But even chemistry provided no magic key to the problems of mineral classification. Faced with ordering the silicates, a problem that had plagued European mineralogists for two decades, Cleaveland confessed his perplexity. The composition of these "earths" was not yet sufficiently understood to be used as a basis for specific or even generic distinctions.

Analysis can indeed inform us what earths are present in these minerals and in what proportions, but it has not yet been able to discover in what manner these earths are here combined, nor to distinguish between those ingredients, which are essential to the composition and those, which are not, and which may in fact be considered as accidentally present.[27]

He speculated that these difficulties arose partly from the imperfection of chemical techniques and partly from ignorance about which of the several ingredients of a "compounded mineral" were the most influential in determining its character. Various substances unessential to the species may have become intermixed with more essential ingredients "from the various and complicated affinities, existing between the several earths and their compounds. Foreign ingredients may thus have been interposed, or essential ingredients made to exist in excess." Thus, Cleaveland echoed the views of Haüy, Brongniart, and Beudant, proceeding no further than they had in solving these puzzles.

Cleaveland did make some progress in distinguishing between rocks and minerals. Under the traditional five Wernerian geological formations he listed the "minerals which fall under the cognizance of geology:"

(1) primitive or primary rocks, among which are granite, gneiss, micaceous slate, and certain limestones, which never contain organic remains, and which occupy the lowest position with respect to the center of the earth;

(2) transition rocks, such as graywacke and certain limestones, which sometimes contain petrifactions;

(3) secondary rocks, such as sandstones, which

[26] Parker Cleaveland, *An Elementary Treatise on Mineralogy and Geology; Being an Introduction to the Study of These Sciences, and Designed for the Use of Pupils—for Persons Attending Lectures on These Subjects—and as a Companion for Travelers in the United States of America* . . . (Boston, 1816), p. 70.
[27] *Ibid.*, pp. 74–75.

abound with organic remains, and which appear to be chiefly mechanical deposits from water;

(4) alluvial deposits, such as clay, sand, and pebbles; and

(5) volcanic productions.

By separating out these "minerals which fall under the cognizance of geology," Cleaveland contributed to the development of a sharper definition of the term *mineral*.

In his fourth chapter, on nomenclature, Cleaveland described the confusion that had resulted from the practice of giving different names to the same mineral. An outstanding example was the mineral called *epidote* by Haüy. This mineral was called *pestazit* by Werner, *thallite* by Lamétherie, *akantione* by Dandrada, *delphinite* by Saussure, *glassy actynolite* by Kirwan, *arendalit* by Karsten, *glasiger strahlstein* by Emmerling, and *la rayonnate vitreuse* by Brochant. To add to the confusion, different writers gave the same name to different minerals. Others invented names, such as Haüy's *grammatite analcime*, and *chabasie*, based on some peculiar property of a mineral. Cleaveland himself used only well-known names, employing the nomenclature of Kirwan and Jameson instead of that of Haüy for earthy minerals whose chemical composition remained in doubt.

Cleaveland's classification of minerals, presented in a tabular view, was a compromise between the systems of Brongniart and Haüy. His first class included Haüy's three orders and another order, uncombined acids, but in all other respects followed Brongniart's plan. (See p. 15.) The resemblance to Brongniart's system was most striking in the treatment of the earths, in which the genera were designated by the combinations of elements and compounds contained in the minerals, for example, "alumine and magnesia" or "silex, alumine, and alkali." Each mineral was described in accordance with a common plan, beginning with its common name followed by its external characters, chemical composition (drawn from Bergman, Kirwan, Klaproth, Vauquelin, or Chenevix), the localities where it was found, its uses, and miscellaneous remarks. "Gypsum," for instance, was the common name of *sulphate of lime*, designated as Species 5 of Genus III of Order III of Class I. This mineral, Cleaveland reported, was often found in amorphous masses and frequently as "crystals, whose primitive form is a four-sided prism, whose bases are parallelograms with angles of 113° 8' and 66° 52'; the sides of the base and the height of the prism are as the numbers 12, 13, 32." Gypsum possessed the power of double refraction, he noted. It was softer than carbonate of lime and could generally be scratched with a finger nail. The specific gravity was usually between 2.26 and 2.31, although Kirwan reported it as being sometimes as low as 1.87. Unless impure it would not effervesce when placed in an acid and was

soluble in 500 times its weight of water. According to Bergman, its chemical composition was "lime 32; sulphuric acid 46; and water 22."

The subspecies of gypsum were (1) selenite, or "foliated sulphate of lime"; (2) gypsum, or "fibrous or granular sulphate of lime, having a compact or earthy texture"; and (3) plaster stone, or plaster of Paris, having present "a foreign ingredient, the carbonate of lime, which entitles it to a distinct notice." In addition, Cleaveland identified six varieties of the second subspecies: (*a*) fibrous gypsum, having parallel fibers; (*b*) granular gypsum, occurring in masses having a granular structure and commonly called alabaster; (*c*) compact gypsum, possessing a fine grain; (*d*) branchy gypsum, a rare variety, whose "branches curled or twisted and collected into little tufts"; (*e*) snowy gypsum, "found in small reniform, or flattened masses, which have the aspect of snow"; and (*f*) earthy gypsum, which results from the "disintegration of some of the preceding varieties."

Next to the carbonate of lime, gypsum was more abundant than any other earthy salt, Cleaveland reported. There were "several distinct formations" found among primitive, transition, and secondary rocks. The oldest formation, or primitive gypsum, was "found resting on or contained within primitive rocks; it is most frequently granular; it is usually white; it is sometimes mixed with mica, talc, feldspar . . . but never with clay, marl, nor does it contain the remains of organic beings." Another formation belonged to transition rocks and was "often associated with graywacke slate." In most cases, though, gypsum was "undoubtedly of secondary or late formation," associated with compact limestone and also with sandstone, and almost always with a strata of clay or marl, with which it usually alternated.

Cleaveland also described the secondary gypsum found near Paris, particularly at Montmartre. In the United States, he observed, gypsum was reported near Abingdon, Virginia, by Seybert; near Baltimore by Hayden; at Poland in Trumbull county, New Connecticut (Ohio), by Silliman; and in abundance in New York by Mitchill, especially "in the vicinity of Cayuga lake, whence in 1812, 6,000 tons were exported to Pennsylvania." It was used as fertilizer for grasses, roots, and greens, although its action on the soil or plant was "not perfectly well explained." It was also employed in the "imitative and ornamental arts," but was "less durable and less valuable than marble." When it lost its water of crystallization, it constituted plaster; mixed with quicklime, it formed a good cement. "The finer kinds of plaster, especially when obtained by calcining selenite, being reduced to a powder and mixed with gum water, are employed for casting statues or busts in molds, for taking impressions of medals, etc." They could also be used for stucco.

Cleaveland faithfully reported the American mineral localities sent to him by his informants, taking care to distinguish between confirmed and conjectural discoveries. In this first edition he reported crystallized green oxide of uranium as "probably . . . observed in Maryland, near Baltimore" and earthy green oxide of uranium as "supposed to exist in *Maryland,* near Baltimore; and in Maine, at Brunswick." Only in his second edition (1822) would he confirm the occurrence of the first of these varieties near Baltimore on the authority of Robert Gilmor, Jr. With respect to the second variety, he reported only that it was "said to exist" there. This was the first recognition of a uranium mineral in the United States.[28]

The final section of Cleaveland's *Treatise,* devoted to the geology of the United States, included a geological map. This was a composite of a geographical map prepared by his publisher, Cummings and Hilliard, and a geological map previously published by Maclure. The principal variation from Maclure's map was that Cleaveland, using Mitchill's account of the geology of Long Island, extended the alluvial deposits there beyond the limits laid down by Maclure.

One aspect of Cleaveland's work that produced criticism at home and praise from an unexpected quarter abroad was his description of basalt. Basalt, Cleaveland wrote, had not been found crystallized. It occurred in large amorphous masses either columnar, tabular, or globular in form. He asserted that basalt was a secondary rock usually found in beds or columns. These are frequently traversed by "rents"; "the beds alternate with those of other minerals or are covered by them." "The recent formation of basalt is clearly indicated by the minerals which it accompanies and especially by its alternation with beds of shell limestone and coal and by the organic remains of shells."

It is extremely doubtful whether any Basalt, strictly speaking, has yet been observed in the *United States;* although it is said to exist on the Stony Mountains. The columnar and prismatic masses, which exist in various parts of the U.S. are undoubtedly a *secondary basaltiform greenstone,* which, in some cases, may be *passing into* Basalt. In all specimens, which the writer has seen,

the eye, especially when assisted by a glass, could discover *feldspar,* constituting one ingredient. . . .

The arguments in favor of the aqueous origin of Basalt seem decidedly to preponderate, although difficulties still remain.[29]

By a curious chance, Johann Wolfgang von Goethe read these lines one evening in June, 1818. Joseph Green Cogswell, who had studied under Cleaveland at Harvard, had met Goethe during a trip to Germany in 1817. Knowing Goethe's intense interest in geology and mineralogy, he thought that his former teacher's *Treatise* would be an appropriate acknowledgement of Goethe's hospitality. The gift arrived in Jena on June 16, 1818, and Goethe stayed up until eleven o'clock reading it. Preferring to learn about an unknown country from the point of view of its geology and geography before studying its history, he found Cleaveland's work fascinating. On reading Cleaveland's statement that basalt probably did not exist in the United States, at least east of the Mississippi River, he was profoundly moved. To Goethe, basalt represented all of the quarrelsome characteristics of mankind. Arguments had raged in scientific journals for decades over the origin of basalt. Found throughout Europe, basalt mirrored for Goethe the internecine struggles and wars that characterized Europe's history. Lacking basalt, he thought, the geology of America would produce a human world less emotional and far less strife-torn than that of Europe. Delighted with this idea, he saw to it that Cleaveland was elected to honorary membership in the Mineralogical Society of Jena. In return he was pleased to receive the second edition of Cleaveland's *Treatise,* sent to him in June, 1823.[30]

In the United States Cleaveland's discussion of basalt was less favorably received. Thomas Cooper was especially critical in his review of Cleaveland's *Treatise* in the *Analectic Magazine.* Cooper ques-

---

[28] The authors are indebted to Professor Clifford Frondel for the following comment: "I am quite certain that the first recognition of a uranium mineral in the United States is due to Gilmor. . . . Gilmor does not mention the locality in his catalogue of Baltimore minerals in the *American Mineralogical Journal,* nor does Tyson in his 1830 paper in the *American Journal of Science* or his 1837 paper in *Journal of the Maryland Academy of Science.* His discovery probably was made between 1814 and 1822. The mineral very likely was autunite, rather than torbernite, and probably was found in the Jones Falls area where the Baltimore gneiss and pegmatite bodies therein have yielded mineral specimens since 1800 at least. . . . His mineral probably was a weathering product of a uranium-bearing niobate-tantalate or silicate present as an accessory mineral in the gneiss or pegmatite. . . . I searched Cleaveland's collection at Bowdoin but did not find specimens."

[29] *Ibid.,* pp. 278–286.

[30] Thomas A. Riley, "Goethe and Parker Cleaveland," *Pub. Modern Language Assn. America* 61 (1952): pp. 350–374. The idea that basalt crystallized from water was advanced by Conrad Gesner in 1565. In the middle of the eighteenth century, Nicholas Desmarest attempted to prove the igneous origin of basalt by showing that there was a connection between the basalt sheets in the Auvergne region of central France and the extinct volcanic craters there. When, in 1786, Abraham Gottlob Werner asserted that basaltic rocks had been crystallized in a primitive ocean, a heated controversy broke out concerning the origin of basalt, the main issue being the relative importance of igneous and aqueous agencies in the formation and alteration of the earth's crust. The proponents of the igneous origin of basalt were called Vulcanists; those favoring its aqueous origin became known as Neptunists. By 1817, when Werner died, most geologists, including his students, had adopted the view that basalt is of igneous origin. Only a few scientists, Cleaveland among them, still adhered to the Neptunist position. Goethe's interest in geology developed as a result of his appointment as supervisor of the reopening of the copper mines in the Ilmenau region where he lived. Like Cleaveland, he was a confirmed Neptunist.

tioned Cleaveland's assumption of the aqueous origin of basalt and his distinguishing it from lava. The French mineralogist Pierre Cordier had recently shown that, when coarsely powdered, fragments of basalt are exactly similar to those of lava and that the same substances are found in both. Cooper also complained that Cleaveland did not distinguish between anthracite and bituminous coal, that he did not give equal space to the Neptunist and Vulcanist theories, and that he made many errors, which the reviewer enumerated, noting the page numbers. In addition, he believed that Cleaveland should not have placed so much confidence in Haüy. An article by M. Methuon in Oken's *Isis* had demonstrated that the laws of crystal growth were quite different from those assigned by Haüy and had gone far towards destroying Haüy's theories. Students of mineralogy needed assistance in the practical identification of minerals, said Cooper, but the knowledge of Haüy's primitive forms would not help them. Cooper did acknowledge, perhaps with tongue in cheek, that Cleaveland's work was a "solid and judicious compilation" from the best of the foreign mineralogists—Haüy, Brongniart, and Jameson—and that it would "go near to supersede the necessity of purchasing the editions he uses of the authors referred to." In conclusion, Cooper thought it a pity that Cleaveland had not used Jameson's recent three-volume edition [published in 1816] nor the information about mineralogy published in foreign journals during the past year and a half.[31]

Cleaveland was incensed. The review hinted strongly at plagiarism, he wrote to Silliman on May 17, 1817. A few days later he wrote again to Silliman, declaring that the reviewer had misunderstood his meaning in some places:

One of these is in regard to the *practical use* of the *primitive form* of crystals. He does not appear to be sufficiently attentive to my plan, mentioned in the preface, of *uniting* the scientific *arrangement* of the French with the *descriptive language* of the Germans, nor to the *distinction*, which is made between characters suitable for *establishing the species*, and those which may be advantageously employed in recognising a mineral, and *referring it* to its *place* in a system, *already formed*. I do not speak of the primitive form as a character, which is to be *extensively* and *frequently* employed, but as one which is important for designating the species, especially in certain doubtful cases, where chemical *analysis* fails. . . .

Admitting that only a small proportion of minerals were found regularly crystallized, Cleaveland nevertheless maintained:

But what I say of the importance of the *crystalline structure* does not refer to the *extent* or *frequency*, with which this character *may* be employed, but to its impor-

tance in those cases, in which it *can be* employed, viz. in crystallized minerals.

He felt especially wounded by Cooper's parting shot. The *Treatise*, he declared, was quite up to date:

On p. iv, I have cited *systematic* writers only; I ought perhaps to have added others, although they are either cited or referred to in the body of the work. Among these, are Geolog. Trans. v. 1—Aiken's Chem. Dictionary, particularly his *supplements*; Count Bournon's Treatise on carbonate of lime &c. &c. I yesterday received Aiken's 2nd edition of his Manual, published in 1815. . . .[32]

Only one substance—fossil copal—mentioned in Aiken's work, Cleaveland noted, was not included in his own work.

Other American reviewers were much more favorably impressed. A review in the *American Monthly Magazine and Critical Review*, possibly written by Samuel Latham Mitchill and signed "K.," praised Cleaveland's book as "auspicious of the advancement of the physical sciences in the United States" and applauded the efforts of Silliman, Bruce, Seybert, Gibbs, Maclure, and Mitchill to establish mineralogy in America. "Professor Cleaveland," the review continued, "has rendered the subject more scientific, by uniting the chemical method with those of Haüy and others. . . . The descriptive mineralogy of Professor Cleaveland is very good." The reviewer was surprised to find no mention of "aerolites" and noted that the only mineral waters described were those at Ballston, Saratoga, and Lebanon. He also mentioned several omissions. The gypsum localities in Pennsylvania and Kentucky were unnoticed, and no localities for pumice were given, although S. L. Mitchill possessed a red specimen from Missouri and black pumice had been found near Hudson City, New York. These localities could be found in the *Medical Repository* of New York, edited by Mitchill. He concluded that: "Upon the whole, we consider this work a valuable acquisition to the science of mineralogy and take pleasure in recommending it to the attention of students and others interested in the subject. . . ."[33]

The reviewer in *The North American Review*, possibly Walter Channing, wrote an extended disquisition on the advanced state of mineralogy in Europe, describing the intense interest created among fashionable circles in England by Davy's lectures on geology and mineralogy at the Royal Institution and contrasting the situation there with that in the United States, "where there is perhaps as much indifference towards his [Cleaveland's] favorite science, as there is ignorance of it." A partisan of the Wernerian method, the reviewer criticized Cleaveland for including too little

[31] Review of Cleaveland's *Elementary Treatise on Mineralogy and Geology, Analectic Magazine* 9 (1817): pp. 301–314. The review was unsigned but generally attributed to Thomas Cooper.

[32] Cleaveland to Silliman, Brunswick, May 22, 1817, Silliman Correspondence, Beinecke Library, Yale University.

[33] "K.," Review of Cleaveland's *Elementary Treatise on Mineralogy and Geology, Amer. Monthly Mag. and Crit. Rev.* 1 (1817): pp. 183–187.

about the shades of color of the various minerals and chided him for classifying minerals on the basis of chemical composition.

But is there not some danger of losing mineralogy as a distinct science, in the theories, and splendid experiments of chemistry. . . . It [the chemical method] has little if any regard to all that is interesting, and beautiful, in the external phenomena of minerals, although these exclusively depend upon their *nature* and component parts. . . . The features of minerals are always indicative of their true composition, for they are its necessary results. We cannot therefore but feel some doubts of the expediency of adopting any method for their arrangement, which is not principally derived from characters peculiar to them, viz. their external and physical characters.

But this review, too, concluded on a positive note:

We acknowledge ourselves very much indebted to Prof. Cleaveland, for the new species with which his work had made us acquainted, and more especially for the new localities of minerals. . . . The author has paid a part of the debt of science we owe to Europe.[34]

In *The New England Journal of Medicine and Surgery,* John Gorham also deplored the lack of attention to science in America but was hopeful that this neglect would cease.

The Americans, it is said, are a reading people, but if the remark be true, it unfortunately happens for the interests of science, that they dip for the most part into the subjects of politics and religion, and rarely trouble themselves with the propositions of philosophy or the descriptions of natural history. This state of things, however, will not long continue.

Cleaveland's *Treatise* would do much to create a taste for science:

Professor Cleaveland has done an acceptable service to the country by his attempt to diffuse the knowledge of this science. It was a task of no small difficulty to accommodate his work to the appetites of different readers, and it required some address to combine accurate description with popular language, and to preserve in a compendium, which should be accessible to most people, a philosophic dress and an air of science.[35]

Gorham then discussed the relative merits of the classificatory systems of Haüy and Werner and contrasted them with the chemical system of Berzelius, concluding that "the advantage is undoubtedly on the side of the chemists." As to Cleaveland's *Treatise,* it was "the production of a man of science, who is capable of writing clearly, and of describing accurately; of one who has brought down the science to the time at which his book was printed; who has not done himself justice in stating the number of works he has con-

sulted, and whose work is recommended to all who are engaged in the pursuit of mineralogy."

Amid these conflicting reviews Cleaveland looked eagerly for one by Benjamin Silliman, his friend and adviser in the long work of planning and writing the *Treatise.* Finally, in 1819, Silliman's review appeared in the first issue of the newly established *American Journal of Science.* Beginning with an overview of the progress of mineralogy in America, Silliman went on to assess the importance of Cleaveland's contribution to the science:

. . . the work of Professor Cleaveland was imminently needed: the science, at large, needed it; and to American mineralogists it was nearly indispensable. It appeared at a very opportune moment. . . . Professor Cleaveland is therefore entitled to our thanks for undertaking this task. . . .

As to Cleaveland's plan of classification:

A happier model could not, in our opinion, be chosen; and we conceive that Professor Cleaveland is perfectly consistent, and perfectly perspicuous, when adopting the chemical composition of minerals as the only proper foundation of arrangement, and of course, rejecting the principle of Mr. Werner, which arranges them upon their external properties, he still adopts his *descriptive* language as far as it answers his purpose.

The "Introduction," Silliman continued, was "at once copious, condensed, and perspicuous," the chapter on the principles of mineral classification "worthy of all praise." "This difficult subject is here discussed with such clearness, comprehensiveness, and candour, as proves the author to be completely master of his subject."

But Silliman was not completely uncritical. He noted that several mineral localities had been omitted and joined "K." in the opinion that meteorites ought to have been included. The carbonate should have been placed first in the lime genus, and quartz first in the class of earthy compounds. But these were minor objections and matters of preference. On the whole, Silliman concluded, "this work does honour to our country and will greatly promote the knowledge of mineralogy and geology."

Discrimination, perspicuity, judicious selection of characters and facts, and a style chaste, manly, and comprehensive are among the characteristics of Professor Cleaveland's performance.[36]

The reception of Cleaveland's *Treatise* abroad was more favorable than it was in the United States. The *Edinburgh Review* found the book "a work of considerable merit," the first to shed extensive light on the mineral resources and localities of America. Preoccupied with this aspect of the *Treatise,* the reviewer overlooked Cleaveland's biases and his classification

[34] Review of Cleaveland's *Elementary Treatise on Mineralogy and Geology, North Amer. Rev.* 5 (1817): pp. 409–429. The review is unsigned; it may have been by Dr. Walter Channing.

[35] John Gorham, Review of Cleaveland's *Elementary Treatise on Mineralogy and Geology, New England Jour. Medicine and Surgery* 6 (1817): pp. 283–297.

[36] Review of Cleaveland's *Elementary Treatise on Mineralogy and Geology, Amer. Jour. Science* 1 (1818–1819): pp. 35–52. Written by the editor, Benjamin Silliman.

scheme, contenting himself with describing what Cleaveland had reported about American ores and minerals. Looking to the future, he concluded:

We look forward with great hopes to the active exertions of our Transatlantic brethren in this interesting field of scientific inquiry; and we shall expect to see the great outline they have traced, filled up by those detailed examinations of particular districts, where the nature and mutual relations of the different rocks have been diligently and accurately studied. . . . We should be glad to hear of the establishment of a Geological Society, to excite the zeal, and unite the labours of the Geologists of America, and to be the organ of communication between them and the rest of the Scientific World.[37]

Distribution of copies of Cleaveland's *Treatise* in Europe was largely the work of Benjamin Vaughan and his brother William in London.

For myself [wrote Benjamin Vaughan], I forget how many copies I sent; but I *know* that two have reached their destination; viz 1 to my friend Sir Joseph Banks, on whose table it will lie, (& become as publicly known as it would be lying here at the Athenaeum) ; 2. to the Geological Society. In consequence of the latter, I have received an order from the Secretaries of the G. S. for 4 copies ; which I have written to Boston to have sent. I cannot tell all that I have sent; but I sent one copy to Mr. Monroe & another to the Philadelphia Society. Another has gone . . . to the Duke of Manchester, Governor of Jamaica. Both my brothers have had copies. . . . I think I sent one to Dugald Stewart; but I forget particulars. I have sent none to France or to the continent of Europe.[38]

In the following month Vaughan wrote that his brother was sending copies to Haüy and Brochant in Paris, to Professor J. F. L. Haussmann in Göttingen, and to the mineralogical societies in Jena, Edinburgh, and Dublin.

The conveyances from London for Paris by Americans are frequent & the two books for Germany will both go to Goettingen, whither my brother often sends. You may always use my brother on these occasions. The books will go in *your* name. I will write to C & H [Cummings and Hilliard] on the subject, & you need think no more about it.[39]

From London, John A. Vaughan wrote Cleaveland that an American gentleman traveling in Saxony had given the great German geologist Abraham Gottlob Werner such a glowing account of Cleaveland's *Treatise* "that the old man was quite cheered with the hope of seeing some consolidated information on his favorite topic from the western regions. It was promised to be sent to him from England but he died shortly after, before his wish was accomplished."[40] Alexander von Humboldt, he added, had borrowed a copy from the Geological Society of London and forgotten to return it.

Mr. Jameson [Professor Robert Jameson of the University of Edinburgh] has expressed his opinion most favorable and you have the Edinburgh Review for further testimony from the north tho' the article was I believe written by Mr. Brande [William Thomas Brande, professor of chemistry at the Royal Institution]. Dr. [Edward] Clarke of Cambridge the min. Professor and noted traveller, uses no other at his lectures and recommends it to all his hearers as the best, and further the Geol. Soc. and individuals have formed or remodeled their collections on your arrangement.[41]

Henry M. Francis of New York confirmed the high regard in which Cleaveland's book was held in London in a letter to his brother Dr. John W. Francis. "This night," he wrote, "I attended the Geological Society and several gentlemen highly extolled the Cleveland [*sic*] Mineralogy commending it as the best work on that subject in the English language." [42] And in the pages of the *North American Review* an American, probably George Ticknor, reported that Professor Johann Haussmann at Göttingen was using Cleaveland's text as the final authority on American mineralogy.

American visitors in Paris brought back equally encouraging reports. One of them wrote Benjamin Silliman that the Abbé Haüy had described Cleaveland's *Treatise* as "the best elementary work on the science extant." The source of this information was probably Robert Gilmor, Jr., then in Paris. In July, 1819, Gilmor wrote to Cleaveland describing the reception of his book:

It was highly approved by all who saw it in France, England, Scotland, & Ireland, and was constantly out of my possession. I lent it to Haüy, Brongniart, Brochant, Bournon, & Lucas, the last of whom made copious extracts from it, probably with a view to improve the next edition of his work on mineralogy.[43]

Two years later Adam Seybert confirmed Gilmor's report of Parisian opinion: "I can assure you, that in Paris it is considered the best elementary work on

[37] Review of William Maclure, *Observations on the Geology of the United States of America* and of Parker Cleaveland, *An Elementary Treatise on Mineralogy and Geology, Edinburgh Rev.* 30 (1818) : pp. 374–388. In a letter to Cleaveland from London, April 16, 1819, John Vaughan attributed this review to William Brande of the Royal Institution.

[38] Benjamin Vaughan to Parker Cleaveland, Hallowell, Maine, September 1, 1817, Cleaveland Correspondence, Bowdoin College Library.

[39] Vaughan to Cleaveland, Hallowell, Maine, October 13, 1817, Cleaveland Correspondence, Bowdoin College Library.

[40] John A. Vaughan to Parker Cleaveland, London, April 16, 1819, Cleaveland Correspondence, Bowdoin College Library.

[41] *Ibid.*

[42] Henry M. Francis to John W. Francis, London, January 2, 1818, Francis Papers, New York Public Library.

[43] Robert Gilmor, Jr. to Parker Cleaveland, July 20, 1819, Cleaveland Correspondence, Bowdoin College Library. See also *North Amer. Rev.* 6 (1817) : pp. 145, 274–276; Adam Seybert to Cleaveland, Philadelphia, August 17, 1821, Cleaveland Correspondence, Bowdoin College Library. Seybert had just returned from Paris.

mineralogy. Haüy speaks in the strongest terms in its favor."

Meanwhile, Cleaveland was experiencing more tangible proofs of the high esteem in which his work was held. In May, 1818, he was notified that he had been elected a member of the American Philosophical Society. Soon afterward came his election as an honorary member of the Geological Society of London. These honors and the general success of Cleaveland's book made a strong impression at Harvard College, which now offered him a teaching position it had been considering for several years. In July, 1820, Cleaveland informed Silliman of the offer and solicited his advice:

I have received proposals to go to Cambridge. My duties would be, for *one term,* one recitation a day, and one lecture on Chemistry or Mineralogy, and for the other *two* terms *three* recitations a day in Philosophy and Mathematics. Very little time is devoted to Mineralogy. Salary $1700.

To induce me to remain here, my duties are to be one recitation and one lecture a day through the year. A medical school has been created by the Legislature, by which it is expected that my salary may be increased $4 or 500.

I have not yet made any determination. What do you think of it? I seem to be doomed to work hard, wherever I am.[44]

Silliman's reply was not encouraging. The salary, he observed, would be barely sufficient "in that expensive & opulent community." The teaching load would not be materially lighter, and the earth sciences were "evidently not regarded at Harvard as they are here or with you." As to fame and prestige: "Your own light probably shines brighter in comparative solitude than it would do if surrounded by numerous other luminaries; you have light enough I am sure however to shine still, but our eyes are more fixed on the solitary morning or evening Jupiter or Venus than when Orion & a hundred other constellations are in view." [45]

---

[44] Cleaveland to Silliman, Brunswick, July 17, 1820, Silliman Correspondence, Beinecke Library, Yale University.

[45] Silliman to Cleaveland, New Haven, August 11, 1820, Cleaveland Correspondence, Bowdoin College Library. Compare Cleaveland's description of his situation at Bowdoin in June, 1816: "The salary of a Professor in Bowd. Coll. is $800, paid quarterly. There are no perquisites whatever. The Boards do not even engage to furnish them houses, or rent. It so happens, that the College own the house, in which I live, and for which I pay $100, annual rent. No alterations, I presume will be made in the salaries for some time. Two or three years since, the Legislature made us a grant of $3000 a year, for ten years, one quarter part of which is to be appropriated in assisting those students, who are indigent. Previous to this grant, the College were incroaching [*sic*] upon their *capital* $1600 annually. Whether the grant will be continued under the *new order* of things, which is likely to take place, I know not. The funds of the College consist chiefly in uncultivated lands, and in bank stock, which at present is not very productive. The charges made by College against each student are for *tuition* $20. a year—for use of Library $1.50 a year, for *room rent* $5.00 a year—amounting to $26.50

On the other hand, Silliman wrote that he would welcome a move that would bring his friend closer to New Haven.

One wonders what the consequences for Cleaveland's career and for the development of the earth sciences at Harvard would have been if he had accepted the offer. Cleaveland was then at the height of his powers. A portrait of him by Thomas Badger about this time shows him seated in a straight chair, delicate hands resting on the arms of the chair, spectacles held in the fingers of the right hand. He wears a dark coat with lace at the neck and a light vest. There is a rock on the table beside him. The curtain behind him is drawn back, disclosing a bookcase. He sits erect, his hair dark and slightly curly. The eyes are large and luminous under dark brows and a high forehead, the mouth firm and resolute. His expression is intelligent and purposeful, suggesting force and a certain nobility of character. A man to be reckoned with, yet it seems doubtful that he could have become the Benjamin Silliman of Harvard, both because he lacked Silliman's training, personality, and family connections and because there was no one in the Harvard College administration to provide the strong support for the development of the earth sciences that President Dwight had provided at Yale. Yet considering Cleaveland's achievements at Bowdoin, it seems possible that he might have accomplished a similar miracle at Harvard. Apparently Silliman thought it unlikely, or he would have urged Cleaveland to take up the challenge. After long consideration, Cleaveland decided to remain at Bowdoin.

In the midst of these exciting events Cleaveland was hard at work on a second edition of the *Treatise.* By the summer of 1818 the first edition of one thousand copies was nearly exhausted, and Cleaveland sent out an announcement of his intention to produce a second edition, adding an appeal for descriptions of mineral localities throughout the United States.

I am preparing a new edition of my Mineralogy, and solicit from you accounts of new Localities, and such other remarks, as may tend to render the work more useful.

I wish to connect with the account of the Locality some brief Geological notice, viz. whether the mineral occurs in veins, or in beds, or is disseminated—the associated minerals—and the *rock,* which contains them. In most cases, the form, structure, and prevailing color of the mineral may be mentioned.

I also wish to obtain as accurate information, as possible, in regard to all minerals explored for *useful* or *ornamental* purposes, such as Nitre, Common Salt, Marble, Marl, Gypsum, Precious Stones, Steatite, Roof Slate, Clays, Pigments, Anthracite, Graphite, Coal, Ores of metals, Chromate of Lead, &c. The quantity of the aforementioned substances *annually* obtained or manufactured,

---

a year. The other small charges made against them merely pass through the hands of the Treasurer to workmen, sweepers, monitor, bellman &c and for fines, & damages."

the *quality,* including the per cent. of metal yielded by ores, and the *price* are particularly requested.

In addition to the foregoing requests, permit me to ask your answers to the following queries, most of which relate to accounts of Localities published by yourself. The number might perhaps be diminished by further examination of the printed papers. But I prefer sending them in full, as they may elicit additional information. [Apparently these queries were added separately for each correspondent].[46]

The response to his appeal showed that Cleaveland's American mineralogical colleagues were as elated as he with the success of the first American treatise on mineralogy and stood ready to supply all the information at their disposal to make the second edition an even greater success. Recently returned from Paris, Robert Gilmor, Jr., sent a long list of suggestions for improving the *Treatise,* "with such other observations on our own minerals as I thought might be acceptable." These included the views of Haüy, Brochant, Brongniart, and Bournon concerning a mineral from Lake Champlain which Colonel Gibbs had sent to Gilmor under the name coccolite but which the Parisian mineralogists pronounced to be a variety of garnet. Another comment concerned a mysterious Baltimore mineral, analyzed by Bruce shortly before his death, that turned emerald green on being heated with borax. According to Gilmor, Bruce "discovered it to be a new & interesting substance & jokingly proposed calling it *Gilmorite.*" Gilmor also reported that his discovery of chromate of iron in certain magnesian rocks from Hoboken, New Jersey, had been confirmed by Thomas Cooper in Philadelphia at his request and that this substance was now being mined near the Bare Hills in the vicinity of Baltimore and exported to Europe. "I sent Dr. [Eric] Bollman at his request two tons to London where he has a manufactory of chromic yellow." Gilmor added that his trip abroad had enabled him to make his own mineral collection

almost compleat in point of species, and it is probably more so than any other in this country though it may be surpassed in the magnitude & splendour of the specimens, which never was an object with me; on the contrary I have always preferred small ones, rarely exceeding three inches square, and paid attention chiefly to their characters being well defined & the chrystals [*sic*] perfect.

"All my larger and handsome specimens," he added, "I have given to Dr. De Butts for our University, with my duplicates, to form the ground work of a larger cabinet." He also offered to send Cleaveland whatever duplicates were needed for his collection in return for specimens of Maine minerals.[47]

From Philadelphia, Cleaveland received a long and cordial letter from Thomas Cooper, who was then hoping for an appointment at the University of Virginia through the efforts of Thomas Jefferson. Cooper informed him that Isaac Lea was about to publish a memoir on minerals in the vicinity of Philadelphia and gave a long list of mineral localities of his own, concluding: "I will announce your intention of publishing a new edition of your mineralogy, in some of the Journals here, whenever you think the first edition is nearly sold off. I am preparing a treatise on geology, but I shall refer to your mineralogy which I have always recommended as my textbook."[48] Other Philadelphians who responded to Cleaveland's inquiries included John Redman Coxe and Charles J. Wister. Coxe reported that he was arranging his personal collection of minerals according to Cleaveland's method of classification and suggested that the second edition of the *Treatise* be issued in two volumes. He also described some minerals in his collection not included in Cleaveland's first edition. Wister had little to communicate except that he had enlisted Solomon W. Conrad and Zacchaeus Collins in Cleaveland's service. "I have discovered no error in your Book," he added, "which I have often recurred to with renew'd pleasure: I think it the very best work on this subject I have ever met with."[49]

New York mineralogists were equally enthusiastic. Mitchill, Torrey, Griscom, Pierce, Eaton, and Edwin James all sent specimens and localities and answered Cleaveland's queries. From his estate at Sunswick, Long Island, Colonel Gibbs wrote that he was on the verge of publishing a sketch of the mineralogy of the United States, containing "a list of all the minerals discovered in the United States of which I could procure sufficient evidence," and promised to send Cleaveland a copy as soon as it was published. The article never appeared, and it is unclear whether Cleaveland ever received the information promised. But Gibbs did answer Cleaveland's specific queries.[50]

Bull. Business Hist. Soc. 17 (1943): pp. 81–91; and "Eric Bollman, Adventurer, Businessman, and Economic Writer," in: *Essays in American Economic History* (New York, 1944).

[48] Thomas Cooper to Parker Cleaveland, Philadelphia, August 3, 1819, Cleaveland Correspondence, Bowdoin College Library.

[49] Charles J. Wister to Parker Cleaveland, Philadelphia, February 1, 1820, Cleaveland Correspondence, Bowdoin College Library. See also Coxe to Cleaveland, Philadelphia, July 23, 1819, in the same collection.

[50] George Gibbs to Parker Cleaveland, Sunswick, Long Island, July 13, 1819, Cleaveland Correspondence, Bowdoin College Library. Again in December, 1819, Gibbs wrote: "I am about publishing a sketch of the mineralogy of the United States in which you will find further particulars on the subject [of iron bloomeries in the United States]." It is to be hoped that the manuscript referred to by Gibbs will eventually be discovered. His letter of July 13 seems to indicate that his list of minerals included "more than one half the known species" in the classification of Haüy.

[46] There is a copy of this circular at the Bowdoin College Library.

[47] Gilmor to Cleaveland, Baltimore, July 20, 1819, and December 24, 1819, Cleaveland Correspondence, Bowdoin College Library. With reference to Dr. Eric Bollman, see Fritz Redlich's two essays: "The Business Activities of Eric Bollman,"

Soon after hearing from Gibbs, Cleaveland received a sixteen-page letter from Samuel L. Mitchill containing answers to the questions Cleaveland had raised about specimens Mitchill had sent earlier. "I perform this service with the greater pleasure," Mitchill wrote, "inasmuch as I find you a fair dealer, and a just estimator of the labours of your contemporaries." Besides answering specific questions, Mitchill passed along information about the projected Cass expedition "from Detroit to or along the Ouisconsin [sic] and beyond the Mississippi through the country of the Sioux and wandering Tatars of the plains north of the Missouri"; news about William Darby, a Louisiana planter and geographer then visiting in New York; an account of a conversation with Augustus Jessup, the Philadelphia geologist assigned to accompany the Long expedition into the Missouri country; the progress of mining ventures at Shawangunk Mountain near Rochester ("Although they failed to enrich their pockets with money they furnished our cabinets with elegant Specimens."); an account of a lead mine in western Virginia given by the area's congressman ("He gave me permission to examine a box of the native ores in the possession of his agent in Baltimore, and to take from it as many pieces as I wished. They are now in the drawer that contains my lead ores."); and a report on the salt prairies in Arkansas country "chiefly related to me by the Osages at the city of Washington, and by several surveyors, hunters and traders who had been there."

A company of the Osages, while on a visit to me, one evening, drew a map of the region on the floor with a piece of chalk. There was a chief delineator, but they, or several of them took the chalk in succession, and consulted & amended untill they concluded the sketch was correct.[51]

These were only a few of the items Mitchill relayed to Cleaveland from his vast stores of information collected through the years with an energy and enthusiasm rivalling that of his fellow Democrat Thomas Jefferson. No wonder Jefferson called him "the Congressional Dictionary"!

Besides his long established correspondents in Baltimore, Philadelphia, New York, Boston, and Albany, Cleaveland could now draw on informants in the more remote parts of the nation. In Cincinnati, Dr. Daniel Drake, the Mitchill of the West, appealed to Cleaveland to contribute specimens for the cabinet of the newly formed Western Museum Society, adding that he and his colleagues had adopted the arrangement of Cleaveland's *Treatise*, "several copies of which are in this place," and would be glad to send Cleaveland specimens in return.

The Trustees of the Cincinnati College have granted apartments in their edifice and several hundred specimens

of minerals, extraneous fossils and marine shells, previously collected by some members of the society, have been handed in and laid upon the shelves.[52]

From Circleville, Ohio, Caleb Atwater sent descriptions of mineral springs, red and yellow ochre, native bismuth, crystallized sulphate of lime, saltpeter, and nitrate of potash in various parts of Ohio and offered to send a geological map of the western country. Finally, with the arrival of Thomas Cooper and Lardner Vanuxem at South Carolina College, Cleaveland had active informants in that region.

With the aid of these widely scattered collaborators, Benjamin Vaughan's assistance in obtaining foreign books, and Silliman's wise advice and growing skill in chemical analysis, Cleaveland made steady progress on the second edition despite "the *slavery* and *perpetual* occupation in College business, to which I am still doomed." In June, 1822, the new edition appeared.

Whereas the first edition had contained 668 pages in one volume, the additional material Cleaveland now added led his publisher to print the second edition in two volumes. The description of the new mineral localities from his many new correspondents accounted for almost all of the increased size, although Cleaveland was careful to include an appendix on meteoric stones, which both Mitchill and Silliman had demanded. Of his changes, Cleaveland wrote in the preface:

The more important of these alterations consist in changing the arrangement of a few of the species. Thus Carbonate of lime is placed first in the genus, Lime, because reference is so often made to it, as a standard of comparison, in describing the other species of the same genus. For a similar reason, Quartz is placed at the beginning of the earthy class.[53]

Similarly, he took note of Silliman's offhand remark in his review that the Tabular View of Minerals should refer to the page numbers on which the individual minerals were described.

A dozen minerals in the earthy class formerly designated as species were now placed under other species, examples being wernerite as a variety of scapolite, allochroite as a garnet, natrolite as a zeolite, and sommite as a nepheline. Similarly, about a dozen new species were added to the earthy class—turquoise, lazulite, petalite, etc. By including the brucite of Gibbs among these, however, Cleaveland inadvertently added to the confusion surrounding that mineral name,

---

[51] Mitchill to Cleaveland, New York, February 13, 1820, Cleaveland Correspondence, Bowdoin College Library.

[52] Daniel Drake to Parker Cleaveland, Cincinnati, July 13, 1819, Cleaveland Correspondence, Bowdoin College Library. See also Caleb Atwater to Cleaveland, Circleville, Ohio, April 5, 1819, in the same collection. In the preface to his second edition Cleaveland lists his collaborators, acknowledging "particular obligations" to Horace Hayden, James Pierce, Silliman, Torrey, and John W. Webster.

[53] Parker Cleaveland, *An Elementary Treatise on Mineralogy and Geology* . . . (2 v., 2nd ed., Boston, 1822) 1: p. viii.

since the brucite listed was actually chondrodite. In the class of metallic ores, iridium, selenium, and cadmium now also made their appearance. The alterations in the new edition, Silliman remarked in a notice in the *American Journal of Science*, entirely "meet our approbation."

The distribution of complimentary copies to individual scientists, societies, and libraries, both in America and in Europe, was much more organized than that of the first edition. The publisher, Cummings and Hilliard, was furnished an initial list of twenty names to whom copies were to be sent, including Haüy, Brongniart, Mohs, Berzelius, and Alexander von Humboldt. This list was augmented over the next two years to more than fifty, one of whom was Alexander I, Czar of Russia, an avid amateur mineralogist, who acquired the great Forster Collection for the Mining Institute in Leningrad in 1802.

The appearance of his second edition marked the last contribution Cleaveland would make to mineralogy in America. Appointed professor of chemistry and materia medica in the Maine Medical School at Brunswick in 1820, he soon found that his new responsibilities, which included collaborating in the surreptitious shipment of cadavers from as far away as Baltimore, severely limited the attention he could give to mineralogy. He hoped, nevertheless, to prepare a third edition of his *Treatise*. The Bowdoin manuscript collection includes a number of letters showing that he was negotiating with his Boston publisher in 1827 for better royalty arrangements. There is also an interleaved second edition, from which it is apparent that Cleaveland intended to make many changes in his introductory section. The handwritten memoranda include a description of Sir David Brewster's optical experiments on crystals, a summary of Mitscherlich's work on isomorphism and polymorphism, and an explanation of Wollaston's theory of spherical atoms. There are few changes in the descriptive section of the work, however, beyond the minerals listed in his first class, although it appears that he continued to read articles on mineralogy in selected foreign journals until 1841.[54] Perhaps, the appearance of James Dwight Dana's *A System of Mineralogy* in 1837 discouraged further efforts, for he was then fifty-seven years old. Or perhaps he became disenchanted with geology and mineralogy. One of his contemporaries wrote that Cleaveland believed that geology

was tending to assume a hostile attitude toward Revelation. According to this source, he confined his instruction in geology to an elucidation of bare facts, and "every added fact in geology increased his confidence in the Mosaic account of creation." Extreme caution about his health prevented Cleaveland from traveling often or for any great distance. Very conservative, he continued to use chemical apparatus of the kind employed by Dalton and Davy until his teaching career finally ended on the day of his death, October 15, 1858.[55]

Despite his eccentricities, Parker Cleaveland had made an important and lasting contribution to the development of the science of mineralogy in the United States. His *Treatise*, particularly the second edition, went far toward describing the mineral resources of the United States from the Atlantic Ocean to the Mississippi River. At the same time his efforts to collect information about American minerals and their localities from every part of the country served to unify and stimulate American mineralogists and to break down the regional barriers that had long separated them. Cleaveland's work remained the leading American mineralogical text until the publication of the third edition of James Dwight Dana's *System of Mineralogy* in 1850.[56] His contribution is commemorated in the name of the mineral *cleavelandite*.

VII.　YALE MOVES TO THE FORE

Before the emergence of Parker Cleaveland and Benjamin Silliman as leading figures in American mineralogy and geology, medical schools in the urban centers of the new republic had provided the main stimulus to the study of the earth sciences, and medical men had been the chief cultivators of these sciences. Cleaveland's work at Bowdoin College broke sharply with this tradition, but it was Yale rather than Bowdoin that took the lead in establishing chemistry, mineralogy, and geology as a permanent part of the liberal arts college curriculum. And it was Benjamin Silliman who brought about this transformation at Yale and

---

[54] Cleaveland Papers, Bowdoin College Library. In the letters from various individuals there are guarded references to the illegal acquisition of cadavers, which were shipped to Brunswick in whiskey barrels. Concerning Cleaveland's plans for issuing a third edition of his *Treatise*, see the notice in *Franklin Jour. and Amer. Mechanics Mag.* **4** (1827): p. 72: "Professor Cleveland [*sic*] is now engaged in preparing for the press, another edition of his Mineralogy, in which the state of this Science, at the time of the publication of the work, will be exhibited, as far as may be practicable."

[55] Leonard Woods, *The Life and Character of Parker Cleaveland* (2nd ed., Brunswick, Maine, 1860).
[56] In a letter to one of the authors, Professor Clifford Frondel comments on the relative merits of Cleaveland's *Treatise* and the early editions of James D. Dana's *System of Mineralogy* as follows: "Hesitant as any one would be to comment disparagingly on J. D. Dana, I would note that the first edition of the *System* was written in 5 or 6 months at a very tender age; it is a rewrite job of Mohs, chiefly, with contributions from a few other sources. Cleaveland's work seems to me to be a more digested and knowledgeable account (of European works, to be sure), better based on chemistry, and in closer touch with the realities of minerals. I suspect that Dana caught up with Cleaveland, in an intellectual sense, along about the 3rd edition of the *System*. Allowing for the difference in time between their respective first editions, and especially in view of the scientific climate that Dana enjoyed at Yale, I feel that Cleaveland did the better job."

made his *alma mater* the training ground for a generation of chemists, mineralogists, and geologists, including James Dwight Dana, Charles U. Shepard, Edward Hitchcock, Amos Eaton, Chester Dewey, Oliver Hubbard, George T. Bowen, Denison Olmsted, and his own son, Benjamin Silliman, Jr.

Yet, strangely enough, Silliman knew nothing about the subjects he was to teach so brilliantly when, on a fine summer day in July, 1801, President Timothy Dwight persuaded him to accept an appointment as professor of chemistry and natural history at Yale College, from which he had graduated only five years earlier. To prepare himself for this task Silliman journeyed to Philadelphia, the leading scientific center in the United States, to learn chemistry from James Woodhouse. Even at this early date Silliman seems to have considered mineralogy a part of his studies, for he took with him a candle box of minerals belonging to Yale College to have them identified. During the winters of 1802–1803 and 1803–1804 he pursued his studies in Philadelphia, where, as we have seen, he made the acquaintance of Adam Seybert and collaborated with Robert Hare in experiments on the fusion of metals and earths with the oxyhydrogen blowpipe. He learned mineralogy as well as chemistry, for his first course of lectures at Yale in the winter of 1804–1805 included "some notices of mineralogy."

In the spring of 1805 Silliman set off for London and Edinburgh, armed with a fund of $10,000 for the purchase of books and apparatus. After visiting Liverpool, London, and Leyden and making trips to the tin mines in Cornwall, he settled in Edinburgh, where he took rooms with John Gorham and another New Englander, John Codman of Dorchester, Massachusetts, later an eminent Congregational minister. In a letter to his niece Mary Noyes, written about three months after his arrival, he described the scene of his studies:

It is a large & beautiful town; it contains 80,000 people, the houses are of polished free stone; many of them are 10 stories high & some of them 14 & even 16; sometimes there are two stories under ground; people live in them all, from the highest to the lowest. The town stands on several hills, with deep vallies between them; over these vallies are handsome stone bridges, connecting one part of the town with another. The hills are covered with houses & in many places the vallies too, so that, as you pass over the bridges, you look down perhaps 100 feet, & see streets, people, houses & bustle & stir, below you, & as you look up, you see the same on the hills. You can easily conceive that all this must look beautifully at night, when the houses are all lighted up, & the lamps burning in the streets; nothing can be more brilliant; you see *illuminated* mountains, *illuminated* vallies, & very high above all the town, on a lofty rock, like the mountains near New Haven, you see Edinburgh Castle, & as you walk in the evening, you hear the French horn, the Bugle Horn, & other martial instruments sounding from the castle, such notes of war, as you my dear, have never heard, in your peaceful retreat . . . & I hope will never hear. For, this country is at present all in

arms, for fear of the French, who you know have for several years, been threatening to invade it. Every where, one meets the soldiers clad in red coats, & with shining armour.[1]

Although there were no separate courses in geology and mineralogy at the University of Edinburgh, these subjects were included "in the midst of the chemical lectures." Robert Jameson, recently returned from his studies with Abraham Werner at Freiberg in Saxony, had not yet begun to lecture, but the Wernerian doctrines were eloquently championed by John Murray, "a well-instructed and zealous advocate of the Wernerian theory," while those of Hutton and Playfair were expounded by John Hope, "an ardent and powerful supporter of the Huttonian or igneous theory." Silliman was challenged by these debates. His geological ideas were "crude and unsettled" when he departed for Europe, he later recalled, and he was "but slowly climbing up the ladder of mineralogy." But his visits to various mining districts in England had whetted his curiosity about the structure and contents of the earth's surface and prepared him to enjoy and profit from his studies in Edinburgh. A reading of Playfair's *Illustrations of the Huttonian Theory of the Earth* and Murray's *A Comparative View of the Huttonian and Neptunian Systems of Geology*, brought Silliman to the conclusion that there was truth in both systems.

I felt greatly relieved when I was excused from attempting to compel myself to believe that porphyry, trap in all its varieties, and even granite, had ever been dissolved in water. I became, therefore, to a certain extent, a Huttonian, and abating that part of the rocks which the igneous theory reclaims as the productions of fire, I remained as much a Wernerian as ever. But I held myself aloof from entire committal to either theory, or to any theory except one derived directly from the facts.[2]

Silliman nearly lost his life in his enthusiasm for collecting minerals. On the morning of March 5, 1806, he spent several hours exploring the terrain at the foot of the cliffs of Salisbury Crag near Edinburgh, "stopping every few minutes to examine the rocks, and freighting my pockets with minerals." He had nearly reached the western end of the cliff when a mass of rock came loose above him and fell directly toward him. Seeing a huge stone column below him, he dashed to its protective shelter, abandoning his cane and his collection of minerals. When the avalanche had passed, he returned to retrieve his treasures, only to be driven back by another landslide. Fortunately for the cause of American science, Silliman survived these adventures in behalf of mineralogy. Whether he retrieved the morning's collection is not recorded.

[1] Benjamin Silliman to Mary Noyes, Edinburgh, February 27, 1806, Silliman Correspondence, Beinecke Library, Yale University.

[2] As quoted in Fisher, *Life of Benjamin Silliman* 1: p. 170.

In May, 1806, Silliman and John Gorham returned to the United States. A half-century later Silliman was still impressed with the importance of this year of study abroad for his scientific career:

In relation to professional improvement [he wrote in his "Reminiscences"] . . . a much higher standard of excellence than I had before seen was presented to me, especially in Edinburgh. Upon that scale I endeavored to form my professional character, to imitate what I saw and heard, and afterwards to introduce such improvements as I might be able to hit upon or invent. It is obvious that, had I rested content with the Philadelphia standard, except what I learned from my early friend, Robert Hare, the chemistry at Yale College would have been comparatively an humble affair. In Mineralogy, my opportunities at home had been very limited. As to geology, the science did not exist among us, except in the minds of a very few individuals, and instruction was not attainable in any public institution. In Edinburgh there were learned and eloquent geologists and lecturers, and ardent and successful explorers; and in that city the great geological conflict between the Wernerian and Huttonian schools elicited a high order of talent and rich resources both in theory and facts. Here my mind was enlightened, interested, and excited to efforts which, through half a century, were sustained and increased. Had I remained at home, I should probably never have reached a high standard of attainment in geology, or given whatever impulse has emanated from New Haven as one of the centres of scientific labor and influence.[3]

Arriving in New Haven on Sunday afternoon, the first day of June, 1806, Silliman was eager to make use of the knowledge, books, and instruments he had acquired during his studies abroad. He spent the summer investigating the geology and mineralogy of the countryside around New Haven and read a memoir about the area before the Connecticut Academy of Arts and Sciences in September. This was published in their *Memoirs* in 1810 and again in a much revised form in the third number of Bruce's *American Mineralogical Journal*. Passing rapidly over the lowland regions as containing "very little which is interesting to a *mineralogist*," Silliman centered his attention on the basaltic escarpments known as East Rock, Pine Rock, and West Rock, comparing them to the Salisbury Crag at Edinburgh. The resemblance of Salisbury Crag to East Rock was striking:

It has the same rude perpendicular columns, the same curvilinear form, and nearly the same extent: It has a similar sloping mass of ruins accumulated at its foot; it fronts the same way; it slopes off with the same easy declivity in the rear: Like the East Rock, it reposes on a bed of red sand stone; and finally, on fracture, the stone presents the same appearance. So far as it has been examined, its chemical characters appear to be the same. It melts in the heat of smith's forge, and, on cooling rapidly, presents the same vitreous slag, which the Scotch whin is known to produce. Hornblende and a white substance softer than quartz, probably feldspar, are the principal ingredients of both. The stone is reckoned among the argillaceous class, by some min-

eralogists, and by others, among the siliceous. The predominant ingredient is certainly silex, or the flinty earth, although when breathed upon, it emits the smell of *clay*, which would induce one to refer it to the argillaceous family.[4]

The Huttonians, Silliman observed, attributed basalt, or "whin stone," to subterranean igneous agencies intruding molten rock into superincumbent strata. In the long course of time these strata were worn and washed away, leaving the more resistant basaltic formations as rocky eminences above the plains. The Wernerians supposed basalt to be a crystalline deposit from a primeval sea. Silliman preferred to reserve judgment on these rival theories of the earth, "where we usually find so much that is visionary, hypothetical or false." He described the terrain around New Haven, giving an accurate account of the physical appearance and mineral composition of the basalts and showing himself more a mineralogist and petrographer than a geologist in his observations. His associates in the Connecticut Academy of Sciences were well impressed with his knowledge of the earth sciences.

Meanwhile Silliman had begun his course on chemistry, interspersing lectures on mineralogy and geology. He had the small mineral collection, chiefly metallic ores, which he had taken to Philadelphia to be identified, as well as a small collection purchased for the College by his brother, and his own acquisitions in Derbyshire, Cornwall, and Edinburgh.

. . . all these things, when arranged, labelled, and described in illustration of the mineral portion of the chemical lectures, served to awaken an interest in the subject

---

[3] *Ibid.* 1 : pp. 195–196.

[4] Benjamin Silliman, "Sketch of the Mineralogy of the Town of New Haven," *Memoirs Conn. Acad. Arts and Sciences* 1, 1 (1810): pp. 83–96. The much revised version of this essay in Bruce's *Journal* 1 (1814): pp. 139–149, was entitled "Mineralogical and Geological Observations on New-Haven and Its Vicinity." This latter paper extends the comparison of basaltic formations to include Mount Tom and Mount Holyoke in Massachusetts and various Old World formations, such as the Giant's Causeway in Ireland. It also extends the description of the vicinity of New Haven—"greenstone and schistose rock not easy to name." Commenting on this paper, Janet Aitken, emeritus professor of geology, University of Connecticut, writes: "In a very general way he describes the succession of schists and near-slates. . . . A considerable section is devoted to descriptions of the mineralogy and lithology of a belt of serpentine and limestone that trends roughly north-south. . . . The description of this complex of greenstone slates, serpentines and limestones is not truly systematic. . . . There is little here that could be classed as analytical. Silliman appears to have been only very mildly interested at this point in interpreting the significance of these rocks and minerals. The descriptions are very uneven, and only those formations he found to be of mineralogical interest receive much attention. . . . Both this and the earlier paper seem designed . . . as general guides for the purpose of encouraging further study." More generally, Professor Aitken observes: ". . . in geology he was essentially a mineralogist-petrographer with a strong focus on igneous rock, basalts in particular." (Letter to the authors.)

of mineralogy, and to produce both aspirations and hope, looking towards a collection which should by-and-by deserve the name of a cabinet. Our own localities in the vicinity of New Haven, containing agates, chalcedonies, phrenite [prehnite], zeolites, marble, and serpentines, were, in the progress of research, not neglected, and the discovery of them in due time excited zeal and afforded pleasure.[5]

About this time Yale was offered an extensive cabinet of minerals owned by a former student at the College, Benjamin D. Perkins. Perkins's father, Elisha Perkins had invented a medical device known as the metallic tractor, consisting of two tadpole-shaped pieces of metal, one of brass and the other of iron or steel, about four inches long. They were applied by placing the two metals in contact, holding them between the thumb and finger, and drawing the points over the afflicted part of the body. The invention sold spectacularly well, and young Perkins was sent to England to market the product there. By the time Silliman, who had known Perkins during his undergraduate days at Yale, arrived in London in 1805 a Perkins Institution had been founded to conduct experiments with metallic tractors. Perkins had joined the Society of Friends, forsworn all vain ostentation, and acquired a substantial collection of minerals, which he brought back with him on his return to the United States about the time that Silliman returned from Edinburgh. The two men met occasionally when Silliman came to New York to see his brother, and he soon found that Perkins was willing not only to display his collection of minerals but also to sell it.

It was arranged [Silliman wrote] in drawers in a case of dark ancient mahogany and when they were displayed successively the appearance was quite captivating. Indeed they captivated me and as we had nothing in the.College comparable to it, it wore [?] upon me, as I called, from time to time, upon my friend Perkins in New York, who expressed a strong wish that *thee* might have this his favorite collection and thus add to the attractions of *thy* lectures. In some departments this cabinet containing about 2000 specimens was rich, the crystallized specimens of calcareous spar were numerous and beautiful, the fluor spars were splendid both in crystals and polished specimens, the quartz family was rich in rock crystals, in amethysts jaspers agates chalcedony & silified [*sic*] wood and in opals. The magnesian family was rich in beautiful serpentines common and noble & in fine asbestos and silky amianthus. Among the tourmalins [*sic*] were some of great beauty. One pair of elongated crystals were so sensitive to electricity that when warmed and one of them was laid in a revolving crochet of the Abby Hauy [*sic,*] attraction and repulsion were very distinctly exhibited, as the other crystal was brought near & then reversed. There was a crystallized diamond with rubies & sapphires. The metallic part of the collection was more limited. There was however gold & silver & platina, native silver in fine filaments ruby & vitreous silver of great richness, splendid crystals of sulphuret of antimony and sulphuret of lead & magnificent ores of Elba iron,

one specimen with the richest rainbow hues & very perfect crystals [iridescent hematite], also a cinnabar of great richness. Such were some of the beauties of the Perkins cabinet and some of them were associated with the names of great men to whom particular specimens had belonged.[6]

Excited by the prospect of acquiring these treasures, Silliman appealed to President Timothy Dwight to allocate $1,000 to purchase the cabinet. Despite the protests of the college treasurer, the purchase was approved, and Silliman hurried to New York to superintend the transfer of the minerals. The drawers with their precious cargo were removed from the mahogany cabinet as the collection was moved by cart to the dock, then by packet to New Haven and carted again to Silliman's living quarters in the Lyceum, where it was to remain for several years. News of the acquisition spread rapidly, and numerous visitors, including the governor of Connecticut, Jonathan Trumbull, came to see the minerals. Elated at this evidence of interest in high quarters, Silliman displayed his mineral treasures to his distinguished visitor, urging the governor to take "the beautiful silky amianthus" in his own hands. "I was then twenty-eight years old," Silliman recalled later, "and confess I was not a little gratified that the devotion of five years in my profession at home and abroad had been so far successful." Little did either Silliman or his guest suspect that scarcely two years hence, Silliman would become Governor Trumbull's son-in-law.

Silliman was now ready to attempt a private course of lectures devoted entirely to mineralogy. In 1807 a class of twenty to twenty-five students of both sexes met in his chambers in the Lyceum to receive instruction in mineralogy for a fee of five dollars, "eventually —as the course was enlarged of six dollars—always excepting poor students to whom no charge was made."

They formed a pleasant group around a centre table on which the minerals were placed in a situation to be distinctly seen by every one, as they were described, and occasionally specimens were sent around in japanned China trays for near inspection, or even for handling, where a character was to be learned in that way as by the soapy feel of steatite or by the gravity of barytes and still more of the metals. Easy chemical experiments were also performed by the blow pipe, by acids and other chemical reagents, by precipitation of metals &c. . . . Besides the knowledge which was thus obtained the familiar conversational communications favored friendly feelings between instructor and pupils producing in some instances permanent friendship.[7]

At this point in Silliman's career, an event occurred which gave him an opportunity to make himself

[5] As quoted in Fisher, *Life of Benjamin Silliman* 1: p. 216.

[6] Benjamin Silliman, "Origin and Progress of Chemistry, Mineralogy and Geology in Yale College with Reminiscences of Personal History" (9 v., MS in Beinecke Library, Yale University) 4: pp. 24–25.
[7] *Ibid.* 4: p. 20.

known to a much wider scientific community. Early in the morning of December 14, 1807, a brilliant meteor exploded over Weston, Connecticut, with a series of deafening reports that were heard fifty miles away, scattering fragments over a wide area. Silliman was in the midst of correcting proofs for his American edition of William Henry's *Epitome of Experimental Chemistry* when the news of this event reached New Haven two or three days later. Silliman dropped his work and hurried with his colleague Professor James Kingsley to the scene of the excitement. There he learned that a great fireball had passed over the town of Weston in a southerly direction about half past six in the morning, undergoing three major explosions in the course of a ten-mile trajectory and raining hot stones to the earth at each explosion. By the time Silliman and Kingsley arrived in Weston a large number of these stones had been found around the town, some of them weighing from twelve to thirty-six pounds although most of the fragments had been shivered to pieces by their impact with the earth. In addition to eyewitness accounts of the meteor's appearance, the violent explosions, and the subsequent discovery of meteorites buried in the earth, Silliman and Kingsley acquired a collection of specimens, the largest of them weighing about six pounds.

They possess every variety of form which might be supposed to arise from fracture with violent force [they reported to the Connecticut Academy of Sciences]. On many of them, and chiefly on the large specimens, may be distinctly perceived portions of the external part of the meteor. It is every where covered with a thick black crust, destitute of splendor, and bounded by portions of the large irregular curve, which seems to have inclosed the meteoric mass: This curve is far from being uniform. It is sometimes depressed with concavities, such as might be produced by pressing a soft and yielding substance. The surface of the crust feels harsh, like the prepared fish skin, or shagreen. It gives sparks with the steel. There are certain portions of the stone covered with the black crust, which appear not to have formed a part of the outside of the meteor, but to have received this coating in the interior parts, in consequence of fissures or cracks, produced probably by the intense heat, to which the body seems to have been subjected. These portions are very uneven, being full of little protuberances. The specific gravity of the stone is 3.6. . . .[8]

Not content with a superficial description of the meteor fragments and of their descent from the skies, Silliman subjected them to chemical analysis. This was no easy task for a young man with little previous experience in the analysis of minerals. Guiding himself by the memoirs of such European chemists as Edward Howard, Louis Nicolas Vauquelin, Martin Klaproth, and Antoine François de Fourcroy, all of

whom had published chemical analyses of meteorites within the past five years, he carefully examined the nature of each precipitate and residue at every stage of the analysis, noting at one stage an unexpected residue:

. . . in some of the experiments with sulphuric acid on this supposed magnesia [Silliman wrote], a white matter, in small quantity, remained undissolved at the bottom of the vessel. It could hardly be silex, and preliminary experiments led me to conclude that no lime was present. Was it accidental, or, was there a small portion of alumine? This white matter, when heated with sulphuric acid and sulphat of potash, did not afford crystals of alum, on evaporation. I have not yet had leisure fully to decide this point, but intend to resume it. The stone had a very slight argillaceous smell, when breathed upon.[9]

Had Silliman been able to establish the presence of alumine in the meteorite, he would have added an eighth substance to the seven previously reported in the scientific literature on meteors: silex, iron, magnesia, nickel, sulphur, chrome, and manganese. In 1808 Georges Sage reported finding substantial quantities of alumine in the Salles meteorite, but Vauquelin, on investigating the matter for the Institut de France, was able to discover only a minuscule amount of that substance in the materials Sage had worked with. Silliman's greatest error was in finding over 50 per cent silex in the sample, but in this he was no farther from the truth than other chemists of his time, as the following table shows. This error was undoubtedly due to the incomplete separation of the enstatite ($MgSiO_3$) from the silica residue of the initial digestion of the meteorite in a mixture of nitric and sulphuric acids. On the other hand, Silliman's findings for nickel oxide, iron oxide, and sulphur were reasonably good.

The Silliman-Kingsley memoir was first published in the *Transactions* of the American Philosophical Society in 1809 and soon afterward was republished in the *Memoirs* of the Connecticut Academy of Arts and Sciences and in several European journals. Silliman was especially pleased to learn that it had been read

---

[8] Benjamin Silliman and James Kingsley, "An Account of the Meteor, Which Burst over Weston in Connecticut, in December 1807, and of the Falling of Stones on that Occasion," *Memoirs Conn. Acad. Arts and Sciences* 1, 1 (1810): pp. 149–150.

[9] *Ibid.*, pp. 153–154. See also Edward Howard, "Experiments and Observations on Certain Stony and Metalline Substances Which at Different Times Are Said to Have Fallen on the Earth; Also on Various Kinds of Native Iron," *Philos. Trans. Roy. Soc. London* 92 (1802): pp. 168–212; Louis Vauquelin, "Sur les pierres dites tombées du ciel," *Annales de Chimie* 45 (1802): pp. 225–245; Martin Klaproth, "Des masses pierreuses et métalliques tombées de l'atmosphère," *Mémoires Akademie der Wissenschaften Berlin* 1803: pp. 37–67; Antoine François de Fourcroy, "Mémoire sur les pierres tombées de l'atmosphère, et spécialement sur celles tombées de l'Aigle, le 6 Floréal, an XII," *Annales Muséum National d'Histoire Naturelle* 3 (1804): pp. 101–112; Louis Vauquelin, "Expériences sur l'aérolite tombé aux environs de Parme, pour y découvrir la présence de l'alumine, annoncée par M. Sage," *Annales de Chimie* 69 (1809): pp. 280–284; Georges Sage, "Description du procédé employé pour déterminer l'existence de l'alumine dans les pierres météoriques," *Journal de Physique* 66 (1808): pp. 460–462.

TABLE 2

COMPARISON OF METEORITE ANALYSES*

Meteorite Analyst Date	Benares Howard 1802	Yorkshire Howard 1802	Ensisheim Fourcroy & Vauquelin 1804	L'Aigle Fourcroy & Vauquelin 1804	Weston	
					Silliman 1809	Mason and Wiik** 1965
SiO_2	50.0	50.0	56.0	53	51.5	36.6
MgO	15.0	24.6	12.0	9	13.0	22.8
Fe_2O_3	34.0	32.0	30.0	36	38.0	38.6
NiO	2.5	1.3	2.4	3	1.5	1.9
S			3.5	2	≈ 1	1.9
CaO			1.4	1		

* See footnote 9, chap. VI and footnote 17, chap. II for citations.

** Mason and Wiik's analysis (1965) is here given only for the constituents reported by the other authors in this table. For the complete analysis by Mason and Wiik, see p. 26.

aloud before the Royal Society of London and the Académie des Sciences in Paris. The Weston meteor, he observed, "was admitted to be one of the most extensive and best attested occurrences of the kind that has happened, and of which a record has been preserved." Thanks to it, his name was now widely known in the scientific community both in the United States and abroad.

During his trip to Philadelphia in January, 1808, at the peak of public interest in the Weston meteor, Silliman seriously considered writing a book about it. Kingsley, then professor of classical languages at Yale, would provide a historical narrative of similar occurrences from antiquity onward, Silliman would supply the chemical and mineralogical account, and Professor Jeremiah Day's theory that meteors were bodies orbiting the earth in the same way that comets revolve about the sun would complete this "Yalensian" work on meteors. The project was eventually abandoned, but Silliman inserted a section on meteors in the American edition of Henry's *Epitome of Experimental Chemistry*, on which he had been working when news of the Weston meteor interrupted him.

Silliman's "Notes" to the American edition of Henry's work also included other matters of mineralogical interest. Part II of Henry's text was devoted to the chemical analysis of minerals and mineral waters, but paid no attention to Haüy's theory of crystal structure and its importance for mineral classification. To remedy this omission partially, Silliman included a brief exposition of Haüy's theory in the "Notes" to the successive American editions, concluding with the observation that:

As it has not been demonstrated, and perhaps, from the nature of the case cannot, be, that crystals are actually formed in this manner, we must regard it as a mathematical hypothesis, depending for proof upon the ample manner in which it explains the phenomena, and the exact correspondence of calculation with fact, the theory having been applied successfully by its author in predicting the chemical constitution of substances from the form of their

crystals, thus anticipating the accurate results of analysis.[10]

Silliman also called attention to the uses of Robert Hare's oxyhydrogen blowpipe, republishing the full account of his own experiments on the fusion of refractory substances with the instrument in the third American edition of Henry's work. In this account, read before the Connecticut Academy of Arts and Sciences and published in their *Memoirs*, Silliman described the arrangement he had used to attach an oxyhydrogen blowpipe to the pneumatic cistern in his laboratory at Yale and reported his success in fusing "primitive earths" and other substances hitherto considered infusible. Of the primitive earths, Lavoisier had been able to fuse only alumine with his oxygen gasometer. Hare had added silex, barytes, and strontianite to the list. Silliman succeeded in fusing glucine, zircon, lime, and magnesia, leaving only yttria (of which he could obtain no specimen) as a primitive earth as yet unfused. Among the other substances exposed to Hare's blowpipe were rock crystal, quartz, gun flint, chalcedony, oriental carnelian, red jasper, beryl, olivine, leucite, chrysoberyl, topaz, corundum, spinel ruby, and steatite. These experiments showed "that there is now, in all probability no body, except some of the combustible ones, which is exempt from the law of fusion by heat." For many years thereafter Silliman's original laboratory research continued these investigations into the fusion of metals, using first Hare's oxyhydrogen blowpipe and later the deflagrator, another of Hare's inventions for producing intense heat.[11]

[10] See Silliman's "Notes" to the first American edition (4th English edition) of William Henry's *An Epitome of Chemistry* (New York, 1808), pp. lxviii ff.; repeated in later editions. Silliman's account of "Meteoric Stones" is in the "Additions" to the 2nd American edition (New York, 1810), pp. lxvii-lxxvii.

[11] Benjamin Silliman, "Experiments on the Fusion of Various Refractory Bodies, by the Compound Blow-Pipe of Mr. Hare," *Memoirs Conn. Acad. Arts and Sciences* 1, 3 (1813):

In the course of these activities, Silliman was forming influential friendships. Sometime in 1805 or 1806 Silliman learned that Colonel George Gibbs of Newport, Rhode Island, had purchased a splendid collection of minerals abroad and had deposited them in Newport before returning to Europe for further travel and study. Gibbs was the son of a wealthy Newport merchant, the second of four generations named George Gibbs. Wishing to see his son follow the mercantile business, the elder Gibbs had sent him as supercargo on a voyage to Canton in 1796. On his return, however, young Gibbs traveled to Europe where he acquired the scientific interests that would dominate the rest of his life. Leaving Paris for Lausanne to improve his health, he studied under the well-known mineralogist Heinrich Struve, professor of chemistry and demonstrator of natural history and formerly inspector general of mines and salt works in the canton of Vaud. Under Struve's guidance Gibbs became passionately interested in mineralogy. At this time, Count Gregorii Razumovsky, a Russian nobleman living in Lausanne, decided to return to Russia and offered for sale his cabinet of minerals, assembled at great expense over many years. Gibbs seized the opportunity and returned to Paris with his handsome acquisition. There he learned of still another mineral cabinet for sale, the cabinet of Gigot d'Orcy, one of the farmers general under Louis XVI, who had spent forty years amassing a collection of four thousand minerals only to lose his head to the guillotine. The purchase of the d'Orcy and Razumovsky collections brought the wealthy young American to the attention of Parisian mineralogists, among whom he made numerous friends. With their assistance and that of the Count de Bournon, a French emigré in London, he made further acquisitions from Saxony, Dauphiny, Vesuvius, Padua, Verona, and the extinct volcanoes of the Rhine.

The Count de Bournon and François Gillet-Laumont, J. F. d'Aubuisson, and other Parisian mineralogists proved to be very valuable friends when Gibbs returned to the United States. They kept him supplied with books, specimens, models, and instruments, and Gibbs, in turn, was extremely generous in his gifts to them and to the École des Mines and the Muséum National d'Histoire Naturelle. Something of the warm feeling that had developed in Gibbs's relations with the mineralogists of Paris is apparent in d'Aubuisson's letter to "son ami le Colonel Gibbs" from Paris in May, 1806:

Milord, For some time Descotils and I have been meaning to thank you for the books which you were so kind as to send us from London, which we have received and which have given us much pleasure, chiefly because they afforded proof that you have not forgotten us, and for which we are very grateful to your lordship. Descotils reckons that you have already analyzed all the minerals of your country; but I say that this is not possible until you have traversed the Allegheny Mountains and examined carefully all the details of their structure and all the circumstances of the superposition of the masses which compose them. Moreover, I want only to see you occupied with Mineralogy; in our corrupt Europe you could do like other people, but now that you are on virgin territory, you must devote yourself only to the mineral kingdom despite its coldness. They tell me that you have declared war on England; I would like to see you seize Canada, but on the condition that you undertake to send a mineralogical description of it.

I am leaving tomorrow for Piedmont, I will reside at the foot of the Alps, I shall examine them high and low; make your domain the Alleghenies, and we shall see whether Europe and America are built with the same stones and in the same manner. Do not forget to make observations on the temperature of the interior of the earth at a depth of 50, 60, and even 100 feet. And observe particularly the position and elevations above sea level of the places where you make them.

Do not forget that it is to us that you must send the account of your expeditions and your observations. Whatever you send me you can always address here at the Council of Mines; M. Gillet will see that it reaches me.

Adieu, Milord, labor diligently in America and then come and enjoy yourself in Europe and tell us your stories. Above all, do not go and get yourself shot by some English bullet. Adieu Your devoted friend.[12]

While Gibbs was still in Europe, Silliman arranged an interview with Gibbs's sister and gained permission to inspect some of the minerals that had been stored in a warehouse next to the Gibbs mansion in Newport. Colonel Gibbs returned from Europe in 1807 and moved into a house on the hill opposite to Old Stone Tower, still a famous Newport landmark. The two men soon became fast friends as they joined in exploring the mineralogy and geology of Rhode Island and in the excitement of unpacking the boxes of minerals Gibbs had purchased in Europe.

I had now acquired a scientific friend and a professional instructor and guide, much to my satisfaction

---

pp. 329–339; reprinted in William Henry, *The Elements of Experimental Chemistry* (2 v., 3rd American from the 6th English ed., Boston, 1814): ("Notes"), note 18, pp. 380–388. For Silliman's further researches on fusion with the compound blowpipe and Hare's deflagrator, see *Philos. Mag. and Jour.* **50** (1817): pp. 106–114; *Amer. Jour. of Science* **3** (1821): pp. 87–91; *ibid.* **5** (1822): pp. 108–112; ibid. **6** (1823): pp. 341–353; *ibid.* **8** (1824): pp. 147–149.

12 J. F. d'Aubuisson to Colonel George Gibbs, Paris, May 21, 1806, Gibbs Correspondence, Newport Historical Society. Original in French. In June of the same year Gillet-Laumont wrote Gibbs to thank him for sending 100 copies of "the work of Leopold von Buch which you had printed," adding: "If in the travels which you plan to make in the interior of America for the advancement of science you could send us some of the productions and especially of your observations, we shall receive them with great pleasure, and we shall let it be known that there is in that part . . . a mineralogist who has undertaken to make known its riches and is engaged therein." Apparently Gibbs was in London preparing to depart for America, for Gillet-Laumont enclosed a letter for the Count de Bournon. Bournon, in turn, gave Gibbs a letter of introduction to Archibald Bruce in New York.

[Silliman wrote], and he appeared equally pleased to find a companion in his scientific sympathies and pursuits, especially in a young man full of zeal, and both willing and desirous to work. There were in Newport no other men that were devotees of science, and therefore we, as regards these pursuits, became intimately associated, and were not long in planning excursions on this picturesque and beautiful island, whose physical features of course depend on its geological structures.[13]

During the same summer Silliman and Gibbs visited Boston, where Gibbs introduced him to Judge John Davis and other Boston scientists and literati, including Gibbs's brother-in-law William Ellery Channing. "The summer was a very profitable one for me in a professional view," Silliman recalled. A year later William Maclure came through New Haven, fresh from his travels in Maine, and Silliman took advantage of the opportunity to meet him. Infidel or no infidel, Maclure was an interesting man and a highly knowledgeable geologist, as Silliman discovered during their geological excursions in the vicinity of New Haven.

From the time when he first inspected Colonel Gibbs's magnificent collections Silliman must have hoped that they might eventually be acquired by Yale, but he seems to have said nothing to Gibbs about it. Gibbs was uncertain what to do with his collection. According to Silliman, he wanted to see it established "in connexion with some public institution or at least in such a position in some city that it could be made available to the promotion of mineralogy and of the connected arts and sciences."

Overtures, more or less explicit, were made to Government to open the collection in the Military Academy at West Point or in the city of Washington, but they were not seconded nor did Boston or Cambridge or perhaps New York or Philadelphia meet the views and feelings of Col. Gibbs and the collections remained at Newport, in great part unopened. My acquaintance with the proprietor was continued and occasional interviews at Newport & New Haven sustained our interest both scientific and personal. I had often expressed a wish to study the cabinet and my willingness to resort to any place for that purpose wherever it might be eventually displayed.[14]

Gibbs eventually decided to place his collection on deposit at Yale. In the winter of 1809–1810 Gibbs visited Silliman in New Haven and announced his willingness to exhibit his collection at Yale if Silliman could arrange proper facilities to display them.[15] De-

lighted at the prospect, Silliman approached President Dwight with a proposal to convert part of the second floor of South Middle Hall into a mineral gallery by knocking down dormitory room partitions to create a room forty by eighteen feet. As usual, President Dwight supported his protégé, and the work of remodeling was begun. The specimens Gibbs had acquired from Gigot d'Orcy's estate and from the Count de Bournon arrived from Newport in 1810, and Gibbs came early in the following year with his black servant Scipio to assist in arranging the specimens on the shelves.

It was a delightful recreation to lift the covers and unroll the specimens which had been so long secluded from view [Silliman recalled] and when we turned out something very superb we could hardly restrain our admiration. To lay the minerals upon the shelves, in the order prescribed by Col. Gibbs, was a pleasure to me, to whom this agreeable duty was assigned. The arrangement adopted was nearly that of the Abby Hauy's Mineralogy.[16]

---

[13] As quoted in Fisher, *Life of Benjamin Silliman* 1: p. 219.

[14] Silliman, "Origin and Progress of Chemistry Mineralogy and Geology . . . ," Beinecke Library, Yale University.

[15] Silliman's recollection (see Fisher, *Life* 1: p. 256) was as follows: "In the winter of 1809–10, Colonel Gibbs, on a journey, called on me in the evening, and, as usual when we met, the conversation turned on the cabinet, and I inquired: 'Have you yet determined where you will open your collection?' To my great surprise he immediately replied: 'I will open it here in Yale College, if you will fit up rooms for its reception.' . . . I lost no time . . . in laying the subject before President Dwight." If this conversation really took place in the winter of 1809–1810, it seems strange that Gibbs should

have written Silliman on March 12, 1811, as follows: "I have no minerals open as I have no room here [in Boston] for that purpose & Bruce has been so lazy as to neglect to send on my last year's collection. I am willing however to open one of my European collections at your College, if you can give it a good situation. For that purpose it would be necessary to have a large room say 20 by 30 ft. with shelves & glass doors, as minerals when in drawers are subject to injury from the curious. The time that I should have them will be undetermined but I shall have & make out a collection for the College. If this meets your wishes, I will send them on when ever you please, and will pass a month in New Haven in the course of the summer to examine & arrange them with you, tho' they may be opened by yourself, on their arrival" (Gibbs Papers, Wisconsin State Historical Society Library). Either Silliman's memory was faulty, or Gibbs communicated his intention to Silliman orally well in advance of writing him on the subject in March, 1811. The latter possibility seems unlikely, however.

[16] Silliman, "Origin and Progress . . . ," 4: p. 53. On page 52 Silliman states: "It was not all one uniform collection but was made of several the chief of which were Nos. 1, 2 & 3 named below & several others were subsidiary.
  1. The Cabinet of Count Razamouski [*sic*] of Russia.
  2. The Cabinet of Gigot D'Orcy of Paris.
  3. Many rare and costly minerals from Count Bournon, an exile from France in London.
  4. A German collection of rock specimens.
  5. Many minerals obtained by purchase or otherwise here & there from Saxony, Dauphiny &c.
  6. Volcanic specimens from Vesuvius & the Phlegrean Fields, from Padua & the extinct volcanoes of the Rhine.
  7. Many geological specimens—miscellaneous including fine ichthyolites from Mount Bolia 30 miles from Verona." A notice in the *Medical Repository* 11 (1808): p. 213 described "Gibbs' Grand Collection of Minerals" as it was in 1808: "The collection of the Count *Razamowsky* consists chiefly of the minerals of the Russian empire. It is particularly rich in gold and copper ores, chromates of lead, the native iron of Pallas, Beryls, Jaspers, &c. The Russian specimens alone are about six thousand in number. The remainder are chiefly German and Swiss. To these Mr. Gibbs has added all the newly discovered minerals, a complete collection of English, Swiss, and Italian specimens, including the ancient marbles, Porphyries [*sic*], &c. the muriates and carbonates of

Silliman's role in arranging the collections was cut short by an explosion during one of his chemical experiments which nearly cost him his eyesight, but the brilliant sight that greeted his eyes when he was able to remove the bandages from them was a well-deserved reward for his labors. Gibbs himself was so well pleased with the display that he ordered the Razumovsky collection, acquired during his studies with Heinrich Struve in Lausanne, to be shipped from Newport. The south end of the second floor of the dormitory was now remodeled to match the north end. By June, 1812, the Razumovsky specimens and others from Italy and Germany were in order on the shelves, and Colonel Gibbs was in town with his bride, the former Laura Wolcott, daughter of the Hon. Oliver Wolcott, for the grand opening of the enlarged display.

At this moment came news that dismayed Federalist New England and placed the projected opening of the new mineral gallery in jeopardy. The Madisonian government in Washington had declared war on Great Britain.

A thrill of painful excitement—an electrical stroke—vibrated through the Continent. . . . The painful topic was . . . not without an important bearing upon the peaceful pursuits of science. The question of course arose in our minds: Shall we proceed to open more treasures in a maritime town, treasures which we cannot remove and which may be destroyed by the vicissitudes of war? We concluded, however, to trust in God and proceed with our work.[17]

The cabinet was opened to the public and soon became a major attraction for sightseers passing through New Haven. Gibbs was delighted at the interest and admiration shown by the steady stream of visitors, including such notables as Daniel Webster, Josiah Quincy, and Harrison Gray Otis. "Trains of ladies graced this hall of science; and thus mute and animated nature acted in unison, in making the cabinet a delightful resort," Silliman recalled.

The fame of the cabinet was now blazoned through the land, and attracted increasing numbers of visitors. This collection doubtless exerted its influence upon the public mind in attracting students to the College, and was regarded as a very valuable as well as brilliant acquisition. The collections were all furnished with catalogues, scientific and popular; and it could not happen that the opening and examining of ten thousand specimens, with a frequent reference to the descriptive catalogues, could fail to give greater extension and precision to my knowledge of the subject. I had become a zealous student of mineralogy and geology, and now felt that the time had

come to present them with more strength and fulness than in former years.[18]

Up to this time Silliman had included lectures on mineralogy and geology in his chemical courses, reserving special attention to mineralogy for his private course conducted in the Lyceum. These lectures had proved very popular, and in 1812 Silliman had undertaken to write them down as the first step toward the composition of "a system of mineralogical knowledge which might answer for my pupils to consult as a text book in their visits to the Cabinet."

This undertaking [he wrote to Cleaveland] has caused me to extend my researches and to endeavour to bring to the view of my pupils the excellences of the French writers as well as to illustrate the splendid collection of Col. Gibbs. Hitherto I have carried the thing on regularly & both delivered the manuscripts as lectures & permitted them to be in the cabinet to be consulted by my pupils.[19]

Now, however, Silliman decided to separate the lectures on mineralogy and geology from his chemical lectures and give a special college course on these subjects while continuing his private lectures on mineralogy. The Perkins collection was moved to the mineral gallery alongside the Gibbs collection, and seats and tables were installed so that the gallery could serve as a lecture hall. Colonel Gibbs, who regularly supplied Silliman with the latest mineralogical and geological books from Europe, also established a prize of one hundred specimens from the duplicates in his collection "to be adjudged to such member of the senior division of the Mineralogical Class in Yale College as his companions of the same division designate by more than one half of the whole number of their votes." In 1811 the prize was awarded to Solomon Baldwin of Brookfield, Connecticut: "During his attendance on the lectures, he discovered the beautiful green marble which he is now working in the vicinity of New Haven." Silliman added a second prize for the senior and junior divisions of the class: a copy of Bruce's *Journal* for five years to the best senior and a free ticket to Silliman's mineralogical lectures for the best junior.[20]

The course was in high demand; Silliman wrote Gibbs early in 1813 that he was expecting to have forty or fifty students in the middle of February. He added that he was looking forward to Gibbs's arrival in New Haven and requested him to bring

---

copper from Chili; the spinel and oriental rubies, of which this is the third complete collection existing. Also, a large geological collection. The whole consists of about twenty thousand specimens. A small part of this collection was opened to amateurs at Rhode-Island, the last summer, and the next, if circumstances permit, the remainder will be exposed."

[17] As quoted in Fisher, *Life of Benjamin Silliman* 1: p. 258.

[18] *Ibid.* 1: p. 259.

[19] Silliman to Cleaveland, New Haven, May 6, 1813, Cleaveland Correspondence, Bowdoin College Library.

[20] See the account of these prizes in *Amer. Mineral. Jour.*, "Intelligence," 1: pp. 269–270. In a letter to Gibbs, New Haven, April 27, 1812, Silliman wrote: "Will you enquire whether a machine to cut & polish minerals can be procured in New York. Our students are very desirous of having one & I am confident they would use it industriously" (Gibbs Correspondence, Newport Historical Society).

from Boston a copy of the first volume of Brochant's *Traité élémentaire de minéralogie*.

. . . I am endeavouring to make out a synopsis for a course of mineralogical lectures that I may do the business better than I have ever done & wish to consult Brochant. Have you seen Lucas' Mineralogy? Is Hauy's new edition out? Is there a 3d volume of Brongniart on Rocks & Geology? [21]

In May he sent Gibbs the manuscript of the syllabus with a request for criticisms. "You will observe one of my great objects has been to make known to my class the excellencies of the French works which are almost unknown to English students," he added.

By 1815 the new arrangements at Yale had been regularized. Two courses of mineralogical lectures, one public and one private, were given concurrently in the spring, followed by a course of lectures on geology in the summer. The classification of minerals used in these courses was detailed by Silliman in a letter to Cleaveland about this time. [22] The classification, omitting entries below the species level, was as follows:

Class I.   Acids Uncombined
  Genus I.   Acids with base of sulphur
    Species: 1. Sulphuric acid   2. Sulphurous acid
  Genus II.
    Species: 1. Muriatic acid
  Genus III.
    Species: 1. Carbonic acid
  Genus IV.
    Species: 1. Boracic acid
Class II.   Salts not metallic
  Order I.   Alkaline Salts
  Genus I.   Salts with basis of potash
    Species: 1. Nitrate of potash
  Genus II.   Salts with basis of ammonia
    Species: 1. Muriate of ammonia 2. Sulphate of
             ammonia
  Genus III.   Salts with basis of soda
    Species: 1. Carbonate 2. Sulphate 3. Borate 4.
             Muriate
  Order II.   Earthy salts
  Genus I.   Salts with basis of barytes
    Species: 1. Sulphate   2. Carbonate of Barytes
  Genus II.   Salts with basis of strontites
    Species: 1. Sulphate   2. Carbonate of strontites
  Genus III.   Salts with basis of lime
    Species: 1. Sulphate of lime 2. Anhydrous sulphate

3. Carbonate 4. Aragonite 5. Fluate of
             lime 6. Phosphate of lime 7. Nitrate of
             lime
  Genus IV.   Salts with basis of alumine
    Species: 1. Sulphate   2. Fluate of alumine & soda
  Genus V.   Salts with basis of magnesia
    Species: 1. Sulphate of Magnesia 2. Borate
Class III.   Earthy compounds
  Genus I.   Silex almost pure
    Species: 1. Quartz 2. Silex almost pure but not
             crystallized nor entirely transparent

"Beyond this, [Silliman added,] I have done little in classification except to form certain natural groups among stones similarly constituted as:

  Group. Stones whose essential constituents appear to
         be silex, alumine & fixed alkali—Porcelanite.
         Obsidian. Pearl Stone. Pumice Stone. Pitch
         Stone. Leucite, Lapis Lazuli. Lepidolite.
         Mica. Andalousite.
  Group. Stones composed of silex, alumine, fixed alkali
         & lime—Feldspar. Jade. Nephritic stone.
         Anestone. Bildstein or Figure Stone Nau-
         ite [?].

Among Silliman's audience, especially at the geological lectures, was his staunch supporter and well wisher President Timothy Dwight, whose project of demonstrating the harmony of science and religion was now being realized by the man he had selected for the task fifteen years earlier. Dwight's career was nearing its end, but he had the satisfaction of knowing that his beloved Yale would be a stronghold of both science and religion under Silliman's capable and pious leadership.

About the same time Silliman began preparing a course of twenty lectures on the chemical analysis of minerals, "that I may do something towards analyzing our own native minerals." He wrote Gibbs indicating his desire to purchase "any instruments or tests of analysis other than those which you wish for your own laboratory." Chemistry was now firmly established at Yale, and Silliman could draw on the assistance of his best private pupils as laboratory assistants eager to learn practical chemistry. These included Denison Olmsted, Amos Eaton, and Edward Hitchcock, all of whom were to make important contributions to geology. With the appearance of Cleaveland's *Elementary Treatise* in 1816 Silliman was no longer dependent on his own unpublished syllabus, since Cleaveland had incorporated materials from the same authors, especially Brongniart and Haüy, on whom Silliman relied. Indeed, Silliman had played an important role in shaping Cleaveland's text. Describing his progress in learning to analyze minerals chemically, Silliman wrote to Cleaveland:

As to my "*beginning* to become the Klaproth of America" I should be satisfied with a much humbler place & even that is probably more than I shall ever attain. I assure you I much desire to become a good analyst. There are a dozen things now that need analyzing & I cannot do one of them any justice. What is the latest period that any thing for your book [second edition] can come into the appendix. I have never yet been able

[21] Silliman to Gibbs, New Haven, January 15, 1813, Gibbs Correspondence, Newport Historical Society. In a letter to Dr. Lyman Spalding, New Haven, March 19, 1813 (Beinecke Library, Yale University), Silliman wrote: ". . . my course of chemical lectures commences the last week in October & continues at the rate of four lectures a week till the middle of July with the interruption of three weeks' vacation in January & the same in May; the fee is 16 dollars to those who have not taken degrees at this College & 12 to those who have. The distinct course on Mineralogy commences immediately after the January vacation & continues to the end of the chemical course—two lectures a week—fee 6 dollars."

[22] Benjamin Silliman to Parker Cleaveland, New Haven, November 6, 1814, Cleaveland Correspondence, Bowdoin College Library.

to examine your green mica & the yellow mineral &c. I am about fitting up a new laboratory with many new accommodations & then I hope to have more spirit in some of these things than now.[23]

The new laboratory he mentioned occupied the second story of a new stone building at Yale along with the mineral collections. The building had been projected in 1819 as a single-story commons building with kitchens in the basement. Silliman suggested that the building would be more imposing with a second story and that this addition could be used to house a chemical laboratory and the mineral collections. President Dwight died in 1817, but his successor, Jeremiah Day, took up the role of "noble counsellor and patron" to Silliman. The new building plans were approved, and in the summer of 1820 Colonel Gibbs was once more on the scene to supervise the arrangement of his collection in its new setting. The operation occupied several weeks, and the result was all that Silliman and his friend had wished:

The spectacle was now splendid; a room 84 feet by 40 and nearly 12 feet high presented its walls entirely covered by cases brilliant with glass and still more brilliant from the effect of the minerals which they contained.
The fixtures for the lectures were also transferred to the new room & here the lectures on mineralogy & geology when [were?] given for a series of years.
The cabinet became now an object of more interest than ever. Its position in a well lighted upper room with sufficient space for visitors, or visiting parties, made it a favorite resort and so it has continued to be to this day 38 years, February 12, 1859.[24]

In 1820 Gibbs's collection was still his personal property. It was not until 1825 that the College purchased the collection which had meant so much to the development of the earth sciences at Yale.[25]

In the midst of these efforts Silliman still found time to carry out two projects that he, Cleaveland and Gibbs had discussed for several years: the founding of a geological society and the launching of a scientific journal. As early as 1812 Cleaveland had written Silliman about the need for a geological society, "if it be true, as has been remarked by some gentlemen, that North America affords facilities for Geological enquiries, superior to any other section of the globe."

We have indeed but few Gentlemen in the U.S. warmly engaged in these pursuits but the number is continually increasing, particularly from among those young gentlemen, who are annually leaving our Colleges. But the *smallness* of the number, so far from discouraging, ought to stimulate us to unite & concentrate our efforts. There ought to be a general Repository both for the Mineralogical & Geological *observations* made in this Country, and for the *specimens* collected. Tis true Dr. Bruce's valuable journal offers itself, as a Repository of this kind in part. But both reflection and experiment teach us, that a *journal* of this kind is not a *sufficient bond* of union. It does not offer a sufficient stimulus to observe and to communicate. It rather solicits, than requires. . . .
Now, Sir, is there not at least a partial remedy for these evils? Ought not a Geological society to be instituted immediately? Does not the state of the science require it? . . . Is there any serious objection to the measure? If not, give me leave to suggest, that a proposal of this kind ought to proceed from yourself, rather than any other gentlemen.[26]

Silliman responded favorably but suggested that Colonel Gibbs would be the best person to formally propose the society. Cleaveland then wrote Silliman about the collapse of an earlier project to form a geological society in Boston and the consequent disaffection between Colonel Gibbs and some of the Bostonians. Besides, he urged, it would be desirable for the initiator of the new society to possess a reputation as a "general Scholar" as well as a mineralogist and geologist. But in the following month Cleaveland broached the subject to Gibbs, advancing the same arguments he had employed with Silliman. "Whether it is yet time to form such a society with a hope of success," he wrote, "you can perhaps better judge than any one, from your frequent journeys and extensive acquaintance with mineralogists in this country." [27]
In the end both Gibbs and Silliman took the initiative. "Colonel G is the real father of the G. S. [American Geological Society]," Silliman wrote to Cleaveland. "I only witness the birth." One suspects, however, that Silliman's prestige and political connections in Hartford had much to do with the passage of an act of the Connecticut legislature constituting George Gibbs, Benjamin Silliman, Parker Cleaveland, J. W. Webster, Robert Hare, Robert Gil-

[23] Silliman to Cleaveland, New Haven, May 1, 1820, Cleaveland Correspondence, Bowdoin College Library. See also Cleaveland's earlier letter to Silliman, July 3, 1819: "In your letter of March 1816, you say, you are about commencing a course of lectures on the *analysis of minerals* to render yourself familiar with this difficult branch and to analyze American minerals. We certainly much need *one* such chemist in the U. States [sic] What have you done, and what can you do in the course of a year or less? I should be extremely glad to publish your analyses of the Necronite of *Hayden*, of the *Carbonate of Magnesia*, both crystallised and pulverulent, of Pierce, of *Phosphate of Iron* of New Jersey, the Chromate of iron of Baltimore, your combinations of *Tungsten* and *Tellurium*, the yellow incrustation on *Molybdena*, which I sent you, &c &c."
[24] Silliman, "Origin and Progress of Chemistry Mineralogy and Geology . . . ," 4: pp. 103–105, Beinecke Library, Yale University.
[25] See Fisher, *Life of Benjamin Silliman* 1: p. 278 ff. The purchase price of $20,000 was raised by an appeal to the alumni and friends of the College. Among the latter group of contributors was Robert Gilmor, Jr.

[26] Cleaveland to Silliman, Brunswick, December 3, 1812, Gratz Collection, Historical Society of Pennsylvania. See also Cleaveland's letter to Silliman May 21, 1813, in the same collection.
[27] Cleaveland to Gibbs, Brunswick, June 19, 1813, Gibbs Correspondence, Newport Historical Society.

mor, Jr. and their associates "a body politic and corporate, by the name of THE AMERICAN GEOLOGICAL SOCIETY" on May 31, 1819. The first meeting of the Society was held on Tuesday, September 7, in the "philosophical room" of Yale College at commencement time and a constitution was adopted providing for one hundred regular members, including a president, eight vice presidents, a recording secretary, three corresponding secretaries, a curator and treasurer, a committee of nomination, and a committee of publication. Meetings were to be held at least four times a year; the Society was to be located "provisionally" at New Haven. William Maclure was elected president, and Gibbs, Silliman, Cleaveland, Stephen Elliott of South Carolina, Robert Gilmor, Jr., Samuel Brown of Kentucky, and Robert Hare vice presidents, with the eighth vice presidency left vacant. John W. Webster, Edward Hitchcock, and Rev. F. C. Schaeffer of New York were made corresponding secretaries, Timothy Dwight Porter recording secretary. The Committee of Nomination consisted of Silliman, Gibbs, Cleaveland, and Hare, the Committee of Publication of Gibbs, Webster, and James Pierce. Invitations to membership were sent to mineralogists and geologists in every part of the country, including Benjamin Vaughan in Hallowell, De Witt Clinton in Albany, and Thomas Jefferson at Monticello. Mitchill's name, for some reason, does not appear on the Record Book of the Society, nor does that of Silvain Godon. Bruce, who would have merited a vice-presidency, had died the previous year. In a letter to Cleaveland, Silliman explained the circumstances surrounding the formation of the Society:

I have just procured from our legislature, now in session, an act of incorporation for our *American* Geological Society. It was thought best by Col Gibbs & others to get the thing done here & done promptly for reasons which whenever I may see you I will explain to you. Your name, Gibbs's, Webster's, Hare's, Gilmor's & my own were all that were inserted in the act but our powers are ample & it is contemplated to have the first meeting at New Haven at our approaching commencement in order to organize the society. We are not particularly desirous of locating at New Haven but are willing it should be at Boston or wherever is thought best. I will send you the act as soon as they are printed off.

I agree with you that after Maclure Gibbs should be named President & so I have told him but he says let it rest for the present.[28]

Despite the hopes of its founders, the American Geological Society was not destined to provide a national forum for the development of mineralogy and geology. Maclure, Gibbs, Amos Eaton, Theodric

Romeyn Beck, Horace Hayden, Chester Dewey, and others were generous in donating books and specimens and arrangements were made to house them in the cabinet at Yale College. Papers were read at some of the meetings and published in the *American Journal of Science*, and Maclure and others came as often as they could for the meetings. But the time was not ripe for scientific meetings on a national scale, and the Society tended to become an adjunct of Silliman's program at Yale.

The last meeting of the Society seems to have taken place on November 17, 1828. The president of the Society, William Maclure, was there, accompanied by his old friend Thomas Cooper, then president of the College of South Carolina. Cooper was "patriarchal and venerable," his manners "mild and conciliating," Silliman later recalled. But Maclure bore the marks of age and infirmity. "The brilliant man whom I first saw twenty years before, had now hoary locks; he stooped as he walked, and an ulcer on his leg made him lame." To Silliman's surprise and delight, the two distinguished visitors attended one of his chemical lectures and called at his house. Unfortunately, the conversation turned to a subject on which Silliman and his guests held divergent views, "the moral relations of science and the expositions it gives of the mind and thoughts of the Creator, as they are recorded in his works." No harsh words were exchanged, however, and the visit ended amicably. Maclure and Cooper departed from New Haven, never to return, and the American Geological Society faded quietly out of existence. The books and specimens accumulated during its brief career remained at Yale.[29]

Although the American Geological Society failed to provide a permanent national forum for the earth sciences, the same cannot be said of Silliman's other creation of this period, the *American Journal of Science*. As in the case of the founding of the Society, the project had been urged on Silliman by Cleaveland and Gibbs for some time before he decided to undertake it. As early as September, 1817, Parker Cleaveland, learning from Silliman that prospects for the continuation of Bruce's *Journal* were poor, wrote in return:

You say that you are afraid that Bruce's Journal is *finally dead;* and express a wish to have it *revived*. One thing is certain. This Journal *must not* be discontinued. The honor of our country and more especially the interests of mineralogy forbid.

Dr. B. is most certainly an able editor, and has done much for the cause of American mineralogy. But, if he has given up his lectures, his attention will be less directed to the science of mineralogy, than it has been; and will not this together with other circumstances, render him not only *willing*, but even *pleased* to have the care of the Journal transferred to other hands.

[28] Silliman to Cleaveland, Hartford, May 31, 1819, Cleaveland Correspondence, Bowdoin College Library. The Minute Book of the American Geological Society is in the Archives Room of the Sterling Memorial Library, Yale University. See also Meisel, *Bibliography of American Natural History* 2: p. 392.

[29] Fisher, *Life of Benjamin Silliman* 1: pp. 285–286.

Now, Sir, if the preceding suppositions can be ascertained to be *facts,* give me leave to say, that the Amer. Min. Journal must be transferred to *New Haven* and Professor Silliman *must consent* to become the Editor. Every thing conspires to these results. I have thought much, and conversed some on this subject and feel great confidence in the correctness of my opinion.

I know you will not object to the *labor* and *responsibility,* when you take into consideration the importance of the object in view. Mineralogy cannot well flourish in this country without a regular and well conducted journal. I could say much, very much, in support of of this position, but to your mind these reasons are obvious & familiar.

The only possible objection to the plan just proposed lies in the mode of the transfer. It is important, that nothing should be done in our little band of mineralogists, that can tend to alienate or divide. Were I well acquainted with Dr B. I would write him at once. Can you not suggest some mode of effecting this business with perfect safety? You may speak to me with implicit confidence, and I promise you, that I will not say or write to any one either *more* or *less,* than you shall dictate. Let me hear from you soon for I begin to wax warmer and warmer on this subject of the journal.[30]

Silliman had been giving the matter serious thought. He informed Cleaveland that he had called on Bruce during a recent visit to New York but had not found him in. Returning home on the steamer *Fulton,* he had encountered Colonel Gibbs, who

seemed to have lost all hopes of Dr B either as to his going on with the journal, or, in fact with any thing else, & very warmly pressed me on the same subject that you have done; friend Griscom has repeatedly done the same & it is certainly proper that I should give it a respectful consideration although I reserve to myself full liberty to say no in the end. My avocations are very numerous, [he added] I can hardly see how I can do justice to such an undertaking even if qualified in other respects which I am far from being vain enough to believe is the fact in an adequate degree.

His questions to Cleaveland show how seriously he had been considering the project:

1. As there is no purely scientific journal in this country would it not be best to have a journal in which mineralogy & geology should be prominent but which should also embrace physical science generally, so as to include zoology, botany, nat. phil. & chemy, maths, astrony & notices of the progress of our arts & the application of principles of science to them; these things to be exhibited in original American papers to the exclusion of any foreign paper in the body of the Journal? Under *intelligence* foreign discoveries might be adequately noticed. Would not such a plan be more likely to concentrate our scientific efforts & to afford adequate materials for a frequent publication?
2. Would men of science in our great cities be willing to contribute to a Journal in N Haven?
3. Would our infidels & light minded folks bear to have geology made to illustrate the truth of the Mosaic history of which I am fully persuaded it demonstrates the truth?
4. How frequent & how large should such a Journal be?

5. If I say no cannot you be persuaded to say yes. I told Col Gibbs you were in my opinion the man.[31]

During the next few months Silliman and Cleaveland continued to correspond about a journal devoted primarily to chemistry, mineralogy, and geology but including "the entire circle of the physical sciences and their applications." By March, 1818, Silliman was in a position to send his friend a prospectus of *The American Journal of Science, More Especially of Mineralogy, Geology, and the Other Branches of Natural History; Including Also Agriculture and the Ornamental as well as Useful Arts.* "I saw Dr. Bruce in November," he added; "he was obviously a broken man but he approved of my plan & authorized me to use his name; but he has since paid the debt of nature as you have doubtless seen in the papers."[32]

Poor Bruce! It was he who had first provided a rallying point and avenue of publication for Americans interested in mineralogy and geology and introduced European standards of mineral analysis and description. With what mingled feelings of regret and satisfaction must he have listened to Silliman's description of the proposed new journal, which would continue his own but seek a wider range of contributors and readers by embracing a wider range of sciences. His own contribution to science was ended, but he had the satisfaction of knowing that the enterprise he had begun would be carried forward by a friend no less capable and devoted to the pursuit of the earth sciences than he had been. Nor would Bruce's own contributions be forgotten. In the third number of the new journal appeared a memoir of Bruce, procured by Colonel Gibbs from Nicholas Bayard, with a portrait of Bruce in the prime of life.[33]

---

[30] Cleaveland to Silliman, Brunswick, September 29, 1817, Beinecke Library, Yale University.

[31] Silliman to Cleaveland, October 6, 1817, Cleaveland Correspondence, Bowdoin College Library. This letter is helpful in establishing the chronology of events. In his short sketch of the history of the *American Journal of Science* in the preface to the fiftieth volume of that journal, Silliman states that his conversation with Colonel Gibbs took place in November, 1817. The excerpts from his "Reminiscences" in Fisher's *Life of Benjamin Silliman* (1: p. 272) place the encounter with Gibbs more generally "in 1817." The letter to Cleaveland would seem to date their meeting definitely in late September or very early October, since Silliman says: "I was at New York last week & called on Dr. B. but without finding him. Col. Gibbs who was with me in the steam boat coming home," etc. It should also be noted that Silliman's historical sketch in volume 50 of the *Journal* makes no reference to his original plan to make geology "illustrate the truth of the Mosaic history," but rather says that "of party, either in politics or religion, there was no trace" in the Journal. He does, however, end by quoting Agassiz to the effect that science traces the thoughts of God.

[32] Silliman to Cleaveland, March 2, 1818, Cleaveland Correspondence, Bowdoin College Library.

[33] "Biographical Notice of the Late Archibald Bruce, M.D. . . . ," *Amer. Jour. of Science* 1 (1819): pp. 299–304. The notice is unsigned, but there is a letter in the Gibbs Correspondence (Folder 4) from E. W. Bibby to Colonel Gibbs, dated New York, September 3, 1818, which says: "Mr. Bayard has promised the notes of our friends life as minutely as he can

The first number of the *American Journal of Science* appeared in July, 1818, and by June, 1819, the entire first volume of four numbers, totaling 448 pages, was completed. Following the proposals outlined in his correspondence with Cleaveland and conversations with Gibbs, Silliman announced in the *Journal's* preface that it would "embrace the Physical Sciences and their application to the Arts and every useful purpose." Although it was designed primarily to publish "original American communications," it would contain occasional selections from foreign journals and notices of scientific progress in other countries. The *Journal* was to be a "rallying point" for American scientific efforts, and Silliman called for the submission of papers in the "three great departments" of natural history—mineralogy, botany, and zoology—in chemistry, natural philosophy and its branches, and in mathematics. He also solicited articles on the applications of science in agriculture, manufacture, and domestic economy; communications on music, sculpture, engraving, and painting; essays on comparative anatomy, physiology, and other branches of medicine; biographical and obituary notices of scientific men; and meteorological registers and reports of agricultural experiments.

The *Journal* published papers or notices in almost all of these areas, but the vast majority of the articles were purely scientific. Mineralogy and geology normally took up a third to one-half of each volume in the early years. Initially Silliman wrote a substantial proportion of the papers or notices on these subjects, but he was soon joined by Edward Hitchcock, Caleb Atwater, Amos Eaton, James Pierce, William Maclure, Colonel Gibbs, and Jeremiah van Rensselaer. Articles describing the mineralogy and geology of recently settled regions came from John H. Kain of Tennessee, who wrote a description of the mineralogy and geology of the northwestern part of Virginia (now West Virginia) and eastern Tennessee; W. B. Stilson of Louisville, who submitted a sketch of the geology and mineralogy of part of Indiana, and Caleb Atwater of Circleville, Ohio, who wrote an essay "On the Prairies and Barrens of the West." Other papers described new mineral localities.

By the end of the first year it was necessary to consider publishing two volumes each year. Volume II contained three numbers, but for several years thereafter there were two volumes per year, each containing two numbers and a combined table of contents and index. By 1822, crystallographers like Gerard Troost and chemists such as Henry Seybert and George Bowen had joined the ranks of contributors. Communications were received from far afield—from Henry R. Schoolcraft in Minnesota on the native copper of Lake Superior and from Louis Bringier in Louisiana on the geology of the lower Mississippi region.

Silliman's insistence that the *Journal* was an American undertaking, "one which patriotic and honorable men can approve," did not prevent a favorable reception across the Atlantic. In the preface to Volume III, Silliman proudly inserted comments by Thomas Thomson, Regius Professor of Chemistry at the University of Glasgow:

I hail it (the *Journal*) as the commencement of American scientific periodical works, and have no doubt from the valuable matter which you have already presented us with, that American will rival the most scientific countries in the old world.[34]

In the preface to Volume IV, Silliman reported that the *Journal* was being sent regularly to subscribers in Sicily, Italy, Switzerland, Germany, France, Sweden, Ireland, Scotland, and England, and occasionally to other European countries.

The *Journal* was a resounding success in every way except financially. By the end of the first year, there were barely 350 subscribers. Some of these had not paid the subscription price, as shown by Silliman's appeal at the end of the first volume for "*punctual payment*, since *this publication is an expensive one.*" The initial publishers were J. Eastburn & Co., Literary Rooms, Broadway, and Howe and Spalding, New Haven, but they soon abandoned the publication, finding it unprofitable. Silliman then made an arrangement with S. Converse of New Haven under which Silliman paid for the paper, printing, and engraving of subsequent volumes, as he constantly reminded his readers. By September, 1822, he was more confident that the journal would survive.

A trial of four years has decided the point, that the American Public will support this Journal [Silliman wrote in the "Preface" to Volume V]. Its pecuniary patronage is now such, that although not a lucrative, it is no longer a hazardous undertaking. It is now also decided, that the intellectual resources of this country are sufficient to afford an unfailing supply of valuable original communications, and that nothing but perseverance and effort are necessary to give perpetuity to the undertaking. The decided and uniform expression of public favour which the Journal has received both at home and abroad, affords the Editor such encouragement, that he cannot hesitate to persevere and he now renews the expression of his thanks to the friends and correspondents of the work, both in Europe and the United States, requesting at the same time a continuance of their friendly influence and efforts.[35]

---

call them to mind and the engraving will be ready on Saturday in time I trust for the next number of the Journal." (Bruce's mother was the daughter of Nicholas Bayard.) Apparently Gibbs transmitted the memoir of Bruce to Silliman in December, 1818, for Silliman, in a letter to Gibbs from New Haven dated December 29, 1818 (Gibbs Correspondence) thanks him for the loan of the Geological Transactions & for the other communications received by Judge John Davis and adds: "I am glad to observe a memoir of Bruce & as the portrait is ready (as I think you informed me) it can appear in No. 3."

[34] *Amer. Jour. Science* 3 (1821): p. viii.
[35] *Ibid.* 5 (1822): "Preface."

But Silliman's trials were not over. After completion of Volume X in 1826, his publisher decided to withdraw, and Silliman had to pay for the extra copies of the previous volumes in the publisher's stock. He now assumed control of the entire project, handling both the business and the editorial problems. The publication became literally what it had long been popularly—"Silliman's Journal."

Reviewing the experiences of these early years in the preface to the fiftieth volume (1847) of the *Journal,* which included a complete index to the first forty-nine volumes, Silliman expressed satisfaction that the *Journal* had become self-supporting, although the circulation had never reached 1,000 paying subscribers and rarely exceeded 700 to 800. Silliman also revealed that the *Journal* had had a large free distribution list both at home and abroad, and that over the years he had presented entire free sets of the work to "infant colleges and to scientific institutions."

His sense of satisfaction was amply justified. The *American Journal of Science* was now firmly established as the leading American scientific journal, and the first to attract wide international attention. Silliman was leaving it, with its future assured, in the capable hands of those he had trained and whom he loved and trusted. Publication would be carried into the next generation by his son, Benjamin Silliman, Jr., whose name had first appeared on the title page as associate °editor in July, 1838, and by his son-in-law James Dwight Dana, who had joined the editorial staff in 1846. Silliman had done more than any other individual to establish the earth sciences on a secure and solid foundation in the United States through his teaching, his training of graduate assistants, his constant labors to enrich the collections and laboratory equipment at Yale, his part in founding the American Geological Society, his public lectures, his textbooks, and his wise and resolute editorship of the *American Journal of Science.*

VIII. MINERALOGY IN 1822

The year 1822 was an important one for mineralogists in both Europe and America. In Europe, it marked the end of the brilliant career of René Just Haüy. For three decades Haüy's crystallographic theory had dominated mineralogical studies at the Museum of Natural History and the School of Mines, but, as has been shown, its validity was now being questioned by younger workers interested in the chemical analysis of minerals and in the electrical and optical properties of crystals as well as their morphology. In America, the second edition of Parker Cleaveland's *Elementary Treatise on Mineralogy and Geology* issued from the press, a fitting symbol of the achievements and limitations of American mineralogy. On the American side of the Atlantic, mineralogy was still in its infancy, preoccupied with describing

and classifying the minerals of the United States in accordance with methods laid down by Werner, Kirwan, Haüy, Vauquelin, Brochant, Thomas Thomson, and others of the older generation in Europe. In the years to come a younger generation of field workers trained by Silliman, Bruce, Troost, and Cleaveland would carry further the mineralogical explorations of Nuttall, Pierce, Godon, Maclure, Seybert, and Mitchill on the American continent, and state and federal geological surveys would provide stimulus and support for the development of professional mineralogists and geologists like Edward Hitchcock, Charles T. Jackson, and Charles U. Shepard. But not until the publication of the third edition of James Dwight Dana's *System of Mineralogy* in 1850 would an American mineralogist of international stature emerge.

But if American mineralogy was immature in 1822 compared with its European counterpart, it had come a long way since the November day twenty years earlier when Benjamin Silliman arrived in Philadelphia with a candle box of minerals which neither he nor any of his colleagues at Yale could identify correctly. Looking back to that time in his review of Cleaveland's *Elementary Treatise* in the first volume of the *American Journal of Science* in 1818, Silliman depicted the sad state of American natural history, especially mineralogy, at the beginning of the century.

Notwithstanding the laudable efforts of a few gentlemen to excite some taste for these subjects, so little had been effected in forming collections, in kindling curiosity, and diffusing information, that . . . it was a matter of extreme difficulty to obtain, *among ourselves,* even the *names* of the most common stones and minerals; and one might inquire earnestly, and long, before he could find any one to identify even *quartz, feldspar,* or *hornblende,* among the simple minerals; or *granite, porphyry,* or *trap,* among the rocks. *We speak from experience,* and well remember with what impatient, but almost despairing curiosity, we eyed the bleak naked ridges, which impended over the valleys and plains that were the scenes of our youthful excursions. In vain did we doubt that the glittering spangles of mica, and the still more alluring brilliancy of pyrites, gave assurance of the existence of precious metals in those substances; or that the cutting of glass by the garnet, and by quartz, proved that these minerals were the diamond; but if they were not precious metals, and if they were not diamonds, we in vain inquired of our companions, and even our teachers, what they were.[1]

By 1822 the situation was entirely different. Devotees of mineralogical science were scattered throughout the United States from Maine to Georgia and westward to Cincinnati, Louisville, Nashville, and the Lake Superior region. In Philadelphia a new generation of mineralogists, inspired and taught by the example of Gerard Troost, William Maclure,

[1] Benjamin Silliman, "Review of *An Elementary Treatise on Mineralogy and Geology* . . . By Parker Cleaveland . . . ," *Amer. Jour. Science* 1 (1818): pp. 36–37.

Adam Seybert, Silvain Godon, and Thomas Cooper, were filling the pages of the *Journal* of the Academy of Natural Sciences with solid essays in descriptive mineralogy. In New York, the pioneering efforts of Samuel Latham Mitchill, David Hosack, Archibald Bruce, and William MacNeven were being seconded by enthusiastic young graduates of the College of Physicians and Surgeons like John Torrey, the Beck brothers, and indefatigable field mineralogists like James Pierce. Albany and Troy were emerging as centers of mineralogical and geological research under the impetus supplied by Amos Eaton and the Becks. The New York Lyceum of Natural History had not yet established its own journal, but its members were among the most frequent contributors to Silliman's *American Journal of Science*. In New England, Parker Cleaveland and Benjamin Silliman had established mineralogy in the undergraduate curricula at Bowdoin and Yale, and Silliman's students were moving out from Yale to do the same at Amherst, Williams, and Dartmouth College, the University of North Carolina, the University of Nashville, and elsewhere. The number of mineralogical publications in the United States was multiplying rapidly. In the decade 1810–1820 more than twice as many such publications appeared as had been published before 1810, and the decade 1820–1830 doubled the output of the previous decade.

This increasing mineralogical activity in America did not go unnoticed in Europe. As we have seen, the Parisian mineralogists were kept posted on American developments by William Maclure, Gerard Troost, Archibald Bruce, Colonel George Gibbs and, above all, by Bruce's *American Mineralogical Journal* and its successor the *American Journal of Science*. Cleaveland's *Elementary Treatise on Mineralogy and Geology* was welcomed in Europe both as a guide to American mineral localities and as a clear and well-informed introduction to the study of mineralogy. European scientific journals, which had paid little attention to American science before 1810, were now beginning to take it into account.

As if to symbolize the growing interest of European scientists in American mineralogical and geological research, there appeared in Hamburg in 1822 a small volume entitled *Beiträge zur Mineralogie and Geologie des Nördlichen Amerika's*, compiled from American authors by the same Heinrich Struve who had been Gibbs's preceptor in mineralogy at Lausanne and had welcomed Archibald Bruce there not long after Gibbs's departure. Apparently Struve's interest in American geology and mineralogy had been stimulated by his contacts with Gibbs and Bruce, for he had continued to keep abreast of American developments through correspondence with Dr. William Meade, the wandering Irish mineralogist who contributed several articles to Bruce's *American Mineralogical Journal*, published

a treatise on the mineral waters of Ballston and Saratoga Springs, and greatly impressed John Torrey with his knowledge of American minerals. From this correspondent, erroneously referred to as "Dr. Thomas Meade of Philadelphia" in Struve's book, Struve had received a collection of American minerals, various American scientific publications, and miscellaneous observations on the minerals of North America, including some comparisons of northern European and North American geology and mineralogy. Delighted with this information and realizing how little of it was available in the German states, Struve decided to publish a volume in German containing some of the mineral localities described in Cleaveland's *Elementary Treatise* and a selection of articles from the *Journal* of the Academy of Natural Sciences of Philadelphia and the *American Journal of Science*.[2]

Struve's summary of the mineral localities listed in Cleaveland's *Treatise* laid special emphasis on the varieties of marble found in various parts of the United States. These were described in some detail, with a list of operating quarries. Then followed seventy-one briefer localities, accompanied by occasional notes contributed by Dr. Meade. The main body of the book, however, was devoted to translations of articles from Silliman's *Journal* and the *Journal* of the Academy of Natural Sciences. From the latter Struve selected Lardner Vanuxem's "Description and Analysis of the Table Spar, from the Vicinity of Willsborough, Lake Champlain," Augustus Jessup's "Geological and Mineralogical Notice of a Portion of the North-Eastern Part of the State of New York," and Gerard Troost's "Description of Some New Crystalline Forms of Phosphate of Lime and Zircon," all published in the Academy's *Journal* in 1821.[3] From Silliman's *Journal* he chose an account of a variety of amber by Troost; John Dickson's "Notices of the Mineralogy and Geology of Parts of South and North Carolina"; Henry R. Schoolcraft's report to the War Department on the copper mines on the southern shore of Lake Superior; accounts of mineral localities by Chester Dewey at Williams College, D. W. Barton in Virginia, and Thomas H. Webb in Providence, Rhode Island; Horace Hayden's "Notice of a Singular Ore of Cobalt and Manganese"; and two letters from Major Joseph Delafield and Dr. John J. Bigsby concerning deposits of sulphate of strontian from Put-in-Bay Island in Lake Erie and the geology of the Great Lakes region. Struve failed to mention, however, that Major Delafield had sent a box of specimens from the Great Lakes area, "chiefly lime stone, full of organic remains," to William Buckland at Oxford Uni-

---

[2] Heinrich Struve, *Beiträge zur Mineralogie und Geologie des Nördlichen Amerika's* (Hamburg, 1822). Struve had not yet seen the second edition of Cleaveland's *Elementary Treatise*.

[3] See *Jour. Acad. Nat. Sciences of Philadelphia* 2, 1 (1821): pp. 182–185, 185–191, 55–58.

versity and that Buckland, struck by the similarity between the fossils of the transition limestones in America and those found in similar strata in Europe, had suggested to his correspondent: "Perhaps the above circumstances of the interesting and important analogy in the formation of Northern Europe and America may be worth communicating to Professor Silliman, for his Journal, which I hope may be continued as ably as it is begun." [4]

Buckland's praise was a welcome support to a jour-

---

[4] See *Amer. Jour. Science* 4 (1822) : pp. 24–25 under "Intelligence and Miscellanies." The notices by Dewey, Barton, Webb, Hayden, Delafield, and Bigsby are from this volume, pp. 274–285. The articles by Dickson, Troost, and Schoolcraft are from 3 (1821) : pp. 1–5, 8–15, and 207–216.

nal that had only just begun to pay for the physical materials and manual labor that went into its production. It was also new evidence that the European scientific community was ready to welcome American geologists and mineralogists into partnership in the exciting work of creating a science of the earth. With the publication of the second edition of Cleaveland's *Elementary Treatise on Mineralogy and Geology* in 1822, the establishment of the *Journal* of the Academy of Natural Sciences of Philadelphia and Silliman's *American Journal of Science,* and the proliferation of courses on chemistry, mineralogy, and geology in liberal arts colleges, the future progress of American mineralogy was assured. The pioneering generation had done its work, and done it well.

www.ingramcontent.com/pod-product-compliance
Lightning Source LLC
Chambersburg PA
CBHW070244230326
41458CB00100B/6028